Energy Vulnerability and National Security

To Ingela
For fuelling my life
without ever suggesting any demand restraint measures

Energy Vulnerability and National Security

The Energy Crises, Domestic Policy Responses and the Logic of Swedish Neutrality

Walter Carlsnaes

Pinter Publishers, London and New York

First published in Great Britain in 1988 by
Pinter Publishers Limited
25 Floral Street, London WC2E 9DS

British Library Cataloguing in Publication Data

A CIP catalogue record for this book is available from the British Library

ISBN 0-86187-988-0

Library of Congress Cataloging-in-Publication Data

Carlsnaes, Walter.
Energy vulnerability and national security.
Bibliography: p.
Includes index.
1. Energy policy—Sweden. 2. Petroleum industry and trade—Government policy—Sweden. 3. Sweden—Foreign relations—1950- . 4. Sweden—National security.
I. Title.
HD9502.S82C37 1988 333.79'09485 88-2540
ISBN 0-86187-988-0

Typeset by Spire Litho Ltd, Salisbury, Wiltshire
Printed by Biddles of Guildford

Contents

Preface

This study represents my first venture into the breezy corridors of 'relevant' research. These were longer and more numerous than first anticipated, however, and in the meantime the energy issue may seem to have lost some of its previous topicality. Although my view on this is obviously *parti pris*, I immediately wish to emphasize that such an assumption has little warrant in the present-day facts governing either the energy market or international politics. Despite the oil glut and ebbing petroleum prices, the energy problem will obviously be with us for a long time to come. Although this study focuses primarily on the so-called oil decade (1973–83), this is one of the more sombre lessons taught by that period.

However, the central focus of this study is not the energy crises as such (on which a number of excellent books exist) but its effects on Swedish foreign policy and national security (which is more virgin territory). Here I must confess that initially the idea of analysing this policy area filled me with considerable misgivings. Was there anything relevant *and* new to be said about a policy whose prime characteristic has been its self-conscious and adamant constancy over a period of many years? I am glad to report that such doubts turned out to be entirely misplaced. From having been one of the more boring topics of polite dinner conversation, this issue has recently—especially since the so-called submarine crisis—shed much of its anaemic hauteur and instead become one of the livelier political acts in Sweden.

This is also my premier stab at policy analysis; and it is primarily as such—and not as a descriptive or explanatory study—that this book should be read. There is, of course, policy analysis and policy analysis. Therefore I have indicated briefly at the end of the first chapter the type of approach favoured here. In so far as the latter part of that chapter also provides a more general outline of the book as a whole, there is no need to pre-empt that discussion here.

While working on this book I have had the pleasure not only of exploring new territory (both epistemological and geographic) but also of getting to know—or getting to know better—a number of people willing to give me of their time and expertise. The following have been particularly forthcoming in this regard: Dudley J. Burton, Melvin A. Conant, David A. Deese, Charles K. Ebinger, Robert J. Lieber, Leon N. Lindberg, Amory B. Lovins, Robert Mabro, Alan S. Manne, William B. Quandt and Louis Turner. In addition,

Bengt Sundelius and Gunnar Sjöstedt have read substantial sections of the manuscript, as has Evert Vedung, who not only is my departmental next-door neighbour but probably knows more about Swedish energy policy than anybody else. Sverker Gustavsson stimulated me to think and write about Swedish neutrality despite my initial scepticism regarding the fruitfulness of such an undertaking. While I am grateful to all of these, obviously none of them should bear any responsibility whatsoever for the inevitable shortcomings of this study.

Sections of the book—or of previous versions—have also benefited from discussions at various international conferences: the 1982 World Congress of the International Political Science Association in Rio de Janeiro; two of the series of annual workshops organized under aegis of the European Consortium for Political Research—those in Salzburg (1984) and Amsterdam (1987); and the 1984 annual meeting of the American Political Science Association in Washington, DC. All of this has been made possible by numerous travel grants from various sources, including the following: the Nordic Co-operation Committee for International Politics, the Siamon Fund and the Wallenberg Fund. The Swedish Institute of International Affairs should also be mentioned in this connection, as well as for generously letting its library acquire many new books suggested by me. Its former director, Åke Sparring, not only encouraged me when I first started this project but also opened my eyes to the fact that Swedish energy policy did not begin with the first oil crisis. But above all I am here grateful to the generosity of the Bank of Sweden Tercentenary Foundation, which not only provided for most of the considerable costs surrounding a project of this kind but also made it possible for me to take a leave of absence for two years to work on this book.

Finally, I would like to make it clear that although my wife, Ingela, is often derelict about putting out lights when the last to leave a room, that particular habit has nothing to do with my dedicating this book to her.

1 Introduction

The year 1973 has come to represent a watershed in the postwar economic vicissitudes of the industrailized world. Cheap oil—the lubricant of prosperity for over two decades—quadrupled in price almost overnight, hitting the pocketbooks and national ledgers of rich and poor countries alike. Although a shift in the balance of competitive advantage between petroleum consumers and producers had begun somewhat earlier, this was also the year when the latter definitely entered the front ranks of the international economic community. This is true in particular for the Organization of Petroleum Producing Countries, the cartel of producers whose acronym—OPEC—quickly became a powerful symbol for the come-uppance and financial *savoir faire* of what had previously been a motley group of mostly poor, small and arid states along the periphery of the West.[1]

The immediate trigger for this price shock was the Yom Kippur War in October 1973, which in its wake had brought a potent mixture of production cuts in the Persian Gulf oil wells and a selective Arab oil embargo. Since the United States prior to the crisis had imported about 25 per cent of its oil from the Middle East, and Europe and Japan between two-thirds and three-quarters, this led to the wild competitive bidding scramble in the Western countries which pushed up the price from $2.70 to $10.46 in less than five months.[2] However, despite the traumatic impact of this crisis, it was virtually over by the spring of 1974, when the Arab embargo against the United States was lifted and production cuts rescinded. Although it was widely assumed that OPEC would after this be unable to balance the production of oil sufficiently to sustain the new market price, this proved to be a vain hope. Thus, although the 1974–5 recession caused demand to dip somewhat, it was back to pre-crises levels by 1976.[3] Indeed, by this time most of the effects of the shock seemed to have worn off entirely.

However, the events leading to the fall of the Shah in early 1979, causing a dramatic reduction in Iranian oil production, once again brought the oil supply issue to the forefront. Although several other OPEC countries increased their production in an attempt to cushion this unintended shortfall, the balance in the world oil market was again rocked as oil importers panicked in the face of anticipated shortages. Due to low stockpile levels following four years of a well-supplied—if expensive—market, this reaction led to disorganized buying on the Rotterdam spot market. Despite an International Energy Agency (IEA) agreement in March 1979 to act collectively to reduce oil demand by 5 per cent, several member countries also sent missions to Arab states to negotiate bilateral supply agreements.[4] As a result OPEC prices increased by 170 per cent over an eighteen-month period, for the first time reaching official levels of over $30 per barrel.[5]

Fortunately, the stockpiling of oil to an all-time high during this period facilitated a more restrained response to the next disruption of oil supply, provoked by the Iraqi attack on Iran in 1980. Sluggish demand for expensive OPEC oil and brimming inventories in most of the importing countries caused prices to rise by no more than 10 per cent.

Since 1980, and despite the continued presence of the Iraq–Iran war, the oil market has not been rocked by any sudden price increases. On the contrary, we are at present enjoying a glut market, which for the first time in its history—in March 1983—forced OPEC to decrease the price of oil. The organization also imposed an overall production ceiling on its members and adopted a prorationing agreement, with Saudi Arabia playing the role of swing producer and market balancer.[6] (Previously OPEC had not functioned as a genuine cartel, since it had tried to control only the price of oil, not the quantity of production.) However, in the face of an increasingly intractible 'free-rider' problem involving a number of OPEC members producing well above their quotas, and several non-members likewise increasing production, the Saudis and their allies more recently changed their strategy. Instead of trying to stabilize the situation, they started a price war by substantially increasing their own production, which has led to price levels comparable (in nominal terms) to those noted just after the first crisis.[7] This development has made some energy experts speak of an impending third energy crisis, following those of 1973–4 and 1979–80.[8] This time, however, the problem is not one of supply disruptions and sudden price increases, but of an over-supply of crude and a significant and sustained decline in the price of oil, leading not only to grave economic and social problems in the producer—rather than consumer—countries but also to uncommonly severe reverberations in the financial (and especially banking) centres of the world.

ECONOMIC AND POLITICAL CONSEQUENCES OF THE OIL CRISES

While it is difficult to trace the underlying transmission mechanisms between oil supply and the international economy, it is nevertheless widely acknowledged that the *economic* consequences of this turbulent decade in the history of the world oil market have been extraordinarily profound, extensive and lasting. The first oil crisis was thus almost immediately followed by a worldwide recession breaking postwar records in the rise of both inflation and unemployment as well as in the decline of the real gross national product.[9] Developments subsequent to the second crisis were even more grave and tangible, if perhaps somewhat less traumatic. Despite some success in reducing inflation and restoring growth by 1978, the Western economies were two years later facing double-digit inflation rates, negative or negligible growth rates and unemployment figures higher even than those following the first crisis.[10] The chief economist of the Organization for Economic Co-operation and Development (OECD) has estimated that as a result of the second oil crisis the economic growth losses for members of this organization were roughly 5 per cent in 1980 and 7.5 per cent in 1981, amounting to income losses during these two years of respectively $350 billion and $570 billion.[11] The figures suggested by the chief economist of the IEA are even higher: more than $1 trillion dollars for a comparable two-year period.[12] For the period 1974–81 as a whole, but referring only to the 'big seven' industrial states, Daniel Yergin has calculated a loss in economic growth of $1.2 trillion.[13]

These enormous economic losses were a consequence not only of the combined effects of higher inflation, higher unemployment and lower real output caused directly by the price shocks, but also of the fact that these developments in turn created pressures in most countries to pursue a combination of contractionary policies—to counter inflation—and expansionary policies aimed at stimulating output and reducing unemployment.[14] The result of these 'paradoxical macroeconomic problems' was a decade of 'stagflation', that is the combined visitation of low growth and high inflation, against which most traditional economic instruments have not proven to be very effective.[15] This is true, in particular, for high unemployment rates, which since the recession in 1974–5 seem to have become a chronic feature of most Western economies.

The *political*—and especially foreign policy—consequences of these economic woes have been equally unsettling. First of all, although it has been customary in the past to think of threats to a nation's security almost exclusively in terms of military challenges arising beyond a nation's borders, the energy crises have forced a reconsideration of this traditional view. It has become clear that interruptions in the flow of critically needed resources can be as detrimental to the balance of international power—and hence national security—as more conventional confrontations such as, for example, the 1962 Cuban missile crisis or the Suez imbroglio in 1956. The fact that crises in connection with resource flows are often a consequence of unintended rather than purposive behaviour in no way alters the significance of this security problem; it anything, it merely complicates its perception and hence retards its resolution.

At the same time it must be acknowledged that although the relationship between national security and energy vulnerability has ostensibly been recognized in the capitals of the West, this recognition has not led to well-coordinated policies—either domestic or international—linking the two dimensions. In the United States, as Henry Nau has demonstrated, energy policies and overall political-military strategies were kept on largely separate tracks throughout the 1970s, while in Europe and Japan the response was mainly a decade of pulling and hauling and an attitude of *suave qui peut*.[16] (It did, however, produce a spate of books, reports and research projects, especially in the United States, where this topic for a good number of years constituted an impressive growth industry, both within and outside academia.) Relations within the Atlantic alliance were especially tense during this period, and have not improved after President Reagan's scuttling of the Carter administration's policies in favour of strict market solutions to the energy issue.

Second, these uncoordinated responses must also be seen in the light of a concomitant decline of America's hegemonic leadership in this issue area. As Robert O. Keohane has argued in his book on the erosion of hegemonic regimes, this development has entailed the virtual dismemberment of the old oil regime and the emergence in its place of a set of new rules, new dominant political actors and, *a fortiori*, of new power relationships.[17] This has not meant, however, that a new system-wide regime has emerged to replace the previous one based on strong American leadership and informal arrangements—instituted in conjunction with Great Britain and the major international oil companies—ensuring Western and Japanese access to Middle Eastern oil at prices which were both low and highly remunerative to the oil majors. The OPEC states have not become hegemonic for the oil system (i.e. able to make and enforce rules), and the IEA is basically an insurance regime, which comes into operation only in the event of oil supply disruptions in order to reduce the risks faced by individual consumer countries. Thus, while the former has not even been able to function as a genuine cartel, the latter continues to be based on a defensive, damage-limitation strategy rather than on a political structure of co-operation and stability.

Thus it should come as no surprise that when the IEA was put to its first real political test, namely, after the fall of the Shah in early 1979, the response was much the same as it had been in 1973: oil users scrambling to ensure supplies irrespective of price.[18] In the absence of a strong hegemonic regime, the problem had become one of collective action in classic game-theoretic terms; and—as in 1973—no individual nation wanted to wind up with the 'sucker's payoff' of oil shortages. Thus each followed the 'defecting' strategy of the Prisoners' Dilemma, with the result that individual 'actors, seeking rationally but in an uncoordinated way to protect themselves, worsened matters for everyone'.[19] Although the response to the mini-crisis in 1980 was somewhat less uncooperative, and the IEA in fact helped to keep the price increase to a minimum, the unsettled situation created by the decline of the old oil regime has not improved. In Robert J. Lieber's phrase, this has meant the continued attraction of the 'unilaterist temptation', and with it protectionist-nationalist tendencies, free-rider problems and generally suboptimal outcomes.[20]

ENERGY VULNERABILITY AND NATIONAL SECURITY: SWEDEN AS A CASE STUDY

The present study is not about the above economic and political implications of the oil crises as such, but about a more particular relationship within this issue area: the link between energy vulnerability and national security. More specifically this query will be pursued in terms of a 'case study' of the Swedish experience during the so-called oil decade rather than in the form of a broad, cross-national study of this nexus.[21] This choice of focus can be briefly justified as follows.

First of all, Sweden represents a fairly typical instance of a highly industrialized state depending exclusively on imports to supply her oil needs. She is in this sense *representative* of the type of state in which the problem of energy vulnerability is of particular relevance to her national security and socioeconomic well-being. Second, however, Sweden also represents a *special instance* of the above class of actors, namely a state which by choice and tradition has forfeited the pros and cons of all forms of alliance commitments in her foreign relations. In the view of many this means that Sweden is particularly susceptible to international pressures—from both the East and the West—aimed at undermining the credibility of her foreign policy actions. Since it has often been argued that such credibility depends in the first instance on the level of perceived congruence between formal foreign policy declarations and actual policies in areas of vital national interest, it is imperative that such states succeed in making credible that they are not only *de jure* but also *de facto* unaligned to—and hence not dependent on, or vulnerable to pressures from—any alliance grouping. Therefore, in so far as the energy issue constitutes a crucial policy area of this kind, the fact that Sweden is committed to an non-aligned foreign policy sets in strong relief (i.e. accentuates and highlights rather than blurs or obscures) the link between energy vulnerability and national security. If we are allowed at this point to relax not only most comparative but also certain conceptual criteria, Sweden can thus be said to be not only a representative case in the above sense but also an almost *ideal-typical* actor for the type of analysis proposed here.

In the following chapter, however, we will leave Sweden aside for the time being and instead restrict ourselves to a series of preparatory discussions consisting of the following: (1) a sketch of the basic structural characteristics of the oil crises of the 1970s; (2) a discussion of the policy options which in principle are available to states facing actual or potential energy crises; and (3) a brief description and assessment of how Western industrialized nations in general have reacted—with or without the aid of the instruments which these options have provided—to the energy situation since the early 1970s. My intention here is the modest one of suggesting an appropriate analytic and empirical background against which Swedish responses to the issue of energy vulnerability can subsequently be described and evaluated.

Chapter 3 will be concerned entirely with the development of Swedish energy policy as a whole during the decade 1973–83, with which we are mainly concerned here. Two main aspects will be discussed in some detail: the 'new' comprehensive energy strategy which came into force during the mid-1970s; and the nuclear energy question, which was the dominant political issue in Sweden during much of the decade considered here. Since not much has been written in English on these topics (especially studies covering the whole period in question), I have provided an analysis in this chapter which is perhaps more detailed than is warranted in the present context. Hence, those readers who are mainly interested in the foreign-policy implications of Swedish energy policy may wish to skim over certain sections of it. Chapter 4 will consider the main characteristics of the second political arena on which our focus will be placed here: Sweden's foreign policy doctrine, and especially the policy of neutrality which has constituted its core in some sense or other for close to two centuries.

It is in the final chapter, however, that these two policy areas in contemporary Swedish politics will for the first time be confronted with each other by way of a critical assessment of the mutual compatibility of their main dimensions. I shall there attempt to make a case—based mainly on conceptual explication and logical analysis—for the claim that given the logic of contemporary Swedish neutrality, the energy vulnerability which continues to characterize Sweden's beholdenness to the international environment remains a fundamental national security problem.

Finally, I would here like to emphasize that what follows is in the form of a policy analysis, or—as two scholars writing on the logic of policy inquiry have called it—a *policy argument*.[22] They have described this form of analysis in terms of the 'attempt to establish good reasons for the adoption of a policy', adding that such argumentation incorporates two kinds of premises—normative and empirical.[23] Here I shall be mainly concerned only with the latter, that is with 'claims that in a given context the actions, policies, and programs associated with a policy will in fact efficiently achieve the goal of the policy'.[24] Thus, in examining the Swedish policy of neutrality in the light of the country's energy vulnerability, my aim is to evaluate the soundness of Swedish energy policy given the premises of her foreign policy. However, although policy arguments of this kind usually conclude with recommendations to act in some specified way, this will not be the case here (except, at some points, by way of implication or occasional asides). Although this is on the face of it, therefore, a critical rather than constructive undertaking, it has its specific place within the larger framework of policy analysis. Charles W. Anderson recently pointed at this function when he wrote the following:

We enter the policy arena not at the beginning but always in mid-current, acting always against the background of experience and precedent, structural expectations, the manifest workability of the going concern. . . . Thus we *reevaluate* policy in terms of neglected considerations or different interpretations of principle. Or we debate whether existing policies are adequate to realize desired values.[25]

This neatly summarizes the rationale underlying this book.

NOTES:

1. As noted in a recent study, the term 'OPEC' has in popular parlance become a shorthand to signify not only the formal members of the organization but also those who, like Mexico or Egypt, typically tend to follow OPEC oil policies. This practice will be followed here. Also, it should be noted that OPEC itself is merely an instrument for the activities of individual states, not an autonomous actor in itself. Indeed, it has very little power of its own. See Paul Kemezis and Ernest J. Wilson III, *The Decade of Energy Policy: Policy Analysis in Oil-Importing Countries* (New York, 1984), p. 2.
2. See the table in Robert J. Lieber, *The Oil Decade: Conflict and Cooperation in the West* (New York, 1983), p. 16.
3. See Dermot Gately, 'OPEC and the Buying-Power Wedge', in James Plummer (ed.) *Energy Vulnerability* (Cambridge, Mass., 1982), p. 38.
4. Wilfred L. Kohl, *After the Second Oil Crisis* (Lexington, 1982), p. 2.
5. See Gately, pp. 37ff; and Robert J. Lieber, 'Cohesion and Disruption in the Western Alliance', in Daniel Yergin and Martin Hillenbrand (eds.), *Global Insecurity: A Strategy for Energy and Economic Renewal* (Boston, 1982), pp. 324ff.
6. See Edward L. Morse, 'An Overview: Gains, Costs and Dilemmas', in Joan Pearce (ed.), *The Third Oil Shock* (London, 1983), pp. 1ff.
7. See Hans Jacob Bull-Berg, *American International Oil Policy: Causal Factors and Impact* (London, 1987), pp. 173f.
8. See, e.g., the essays in Pearce.
9. See G.F. Ray, 'Europe's Farewell to Full Employment', in Yergin and Hillenbrand, pp. 217f.
10. For comparisons in terms of GDP growth, consumer prices and unemployment rates between the United States, Canada, Japan and the major European countries, expressed on a yearly basis between 1965 and 1981, see Yergin and Hillenbrand, pp. 377f. See also Alvin L. Alm, 'Introduction: Managing Oil Shocks', in Alvin L. Alm and Robert J. Weiner (eds.), *Oil Shock: Policy Response and Implementation* (Cambridge, Mass., 1984), pp. 1ff; and Barry B. Hughes, 'The First Two Oil Shocks: Policy Response and Effectiveness', *Policy Studies Review*, vol. 5, no. 4 (May 1986), pp. 722ff.
11. These figures are reported by Alm, p. 7.
12. See Lieber, p. 4. These figures refer to both direct and indirect effects as well as to the economic and monetary policies adopted in specific response to the second crisis.
13. See Daniel Yergin, 'Crisis and Adjustment: An Overview', in Yergin and Hillenbrand, pp. 5ff and ibid., pp. 377f.
14. See Robert S. Pindyck and Julio J. Rothenburg, 'Energy Shocks and the Macroeconomy', in Alm and Weiner, pp. 112ff.

15. Yergin, p. 3. The cited characterization is G. John Ikenberry's, in 'The Irony of State Strength: Comparative Responses to the Oil Shocks in the 1970s', *International Organization*, vol. 40 (Winter 1986), p. 109.
16. See Henry Nau 'The Evolution of US Foreign Policy in Energy: From Alliance Politics to Politics as Usual', in Robert M. Lawrence and Martin O. Heisler (eds.), *International Energy Policy* (Lexington, 1980), pp. 37–64; and Lieber, pp. 1ff.
17. Robert O. Keohane, *Beyond Hegemony: Cooperation and Discord in the World Political Economy* (Princeton, 1984), p. 182.
18. Ibid., p. 226.
19. Ibid., pp. 223, 231.
20. See Lieber, pp. 7ff. These specific aspects will be discussed in more detail in the final chapter.
21. The term 'case study' has here been given inverted commas to pre-empt any and all arguments that what follows is not a 'real' case study, that is, a case with explicit comparative ramifications. I am well aware of the essential methodological implications of the term (hence the inverted commas); but the sceptical reader is nevertheless welcome to consult Chapter 3 of my book on *Ideology and Foreign Policy: Problems of Comparative Conceptualization* (Oxford/New York, 1986–7), especially pp. 77ff.
22. David C. Paris and James F. Reynolds, *The Logic of Policy Inquiry* (New York, 1983), pp. 3ff.
23. Ibid., p. 3.
24. Ibid.
25. Charles W. Anderson, 'Political Philosophy, Practical Reasons, and Policy Analysis', in Frank Fischer and John Forester (eds.), *Confronting Values in Policy Analysis: The Politics of Criteria* (Newbury Park, 1987), p. 33.

2 The Energy Crises and International Responses, 1973–1983

CONCEPTUAL ISSUES AND THE STRUCTURE OF THE PROBLEM

Before proceeding to a discussion of some of the structural characteristics of the past oil crises, a brief excusion into the terminology of this analysis is essential. For the time being I will restrict myself to three concepts of fundamental relevance to the ensuing analysis, namely the notions of vulnerability, national security and energy disruption.

Although the concept of *vulnerability* is anything but straightforward, I shall here not go into the details of the widespread discussions which have surrounded it (and its terminological cousins) since it first came into vogue in the late 1960s. Rather, I shall forthwith accept a conceptualization which, according to David A. Baldwin's analysis of the history of the term, has roots running back all the way to Malthus, Adam Smith, Rousseau, Montesquieu, and Machiavelli's *The Prince*, and which during this century has been espoused by, amongst others, Sir Norman Angell, Francis Delaisi and Ramsey Muir (all writing prior to the Second World War), as well as, more recently, by Albert O. Hirschman, Kenneth N. Waltz and Baldwin himself.[1] This conception is exclusively associated with the notion of 'vulnerability interdependence', and does hence *not* incorporate 'sensitivity interdependence', with which the former concept has so often been paired following the publication of Robert O. Keohane and Joseph S. Nye's influential book on *Power and Interdependence*.[2] This summary ejection of the notion of 'sensitivity' from the conceptual confines of 'interdependence' has been given the following forceful justification by Baldwin:

> The price of caviar in the U.S. may be 'sensitive' to price changes in other countries, but no one has suggested that America is 'dependent' on caviar. . . . In sum, there is a distinction between drug users and drug addicts, between drinkers and alcoholics, between being sensitive to others and being dependent on them, between influence in general and dependence as a special kind of influence. It is a distinction that has been recognized and understood by scholars and laymen alike for centuries. It is a distinction that the concept of 'sensitivity interdependence' blurs, but which is captured with precision and parsimony by the Hirschman–Waltz concept of dependence.[3]

This latter concept of dependence—and, *a fortiori*, of interdependence, defined as mutual dependence—refers to the opportunity costs of severing a relationship, in distinction to the notion of mere contingency implied by the concept of sensitivity.[4] More specifically, the claim endorsed here is that it is the cost of disengagement from a given relationship which constitutes the pertinent criterion and measure of dependency. And since 'dependency', 'dependence' and 'vulnerability' will here be used synonymously, this means that the higher such a cost, the higher also the level of vulnerability.[5] Clearly the phenomenon thus defined has little to do with 'a condition in which anything that happens anywhere in the world may affect somebody, or everybody, elsewhere', to quote Waltz's derisive characterization of 'sensitivity interdependence'. Or as he has also argued against such a basically economistic interpretation of this concept:

> Interdependence as sensitivity . . . entails little vulnerability. The more automatically, the more quickly, and the more smoothly factor costs adjust, the slighter the political consequences become. . . . To understand the foreign-policy implications of high or of low interdependence requires concentration on the politics of international economics, not on the economics of international politics.[6]

In short, since 'sensitivity interdependence' as usually conceived is essentially an economist's definition in market terms, the linkages it denotes entail little vulnerability in the sense adduced above and should hence not, according to the view presented here, be conceptually subsumed by the latter notion.

Our focus here will thus not be on the 'normal'—and exceedingly beneficial—conditions of external beholdenness (or sensitivity) which were prevalent during the first decades following the Second World War, when oil imports into most industrialized countries were very substantial but at the same time both cheap and readily available. Rather, we shall discuss the national security implications of a situation when, in the event of supply interruptions, oil import levels are too high to be handled by remedial measures at modest or acceptable cost levels.

As will be seen below, these situations typically arise when states do not dispose of multiple sources of oil, sufficient stockpiles, spare domestic production capacity, or similar practicable options. The reduction of vulnerability is thus not a question of reducing external beholdenness *per se*, but of having recourse to appropriate policy options as insurance against potential damage from interrupted supply. Although this may sound paradoxical, it is thus in principle possible for a state to remain highly sensitive to external energy supplies while at the same time *reducing* its vulnerability to supply interruptions. This would be the case, for example, if a state consciously pursues a policy of augmenting its oil stocks during periods of softening prices. Similarly, a country which has acquired this form of insurance would be less vulnerable than one which, although less sensitive in terms of the proportion of energy that it imports, has failed to implement such measures.

Indeed, the logic of this distinction is such that a zero sensitivity to external resources—ostensibly Richard Nixon's goal after the first oil crisis—is by no means *a priori* the most rational course of action in the pursuit of an optimal energy strategy. As long as imported oil is cheaper than domestic sources of energy, it does not pay to dispense with the former in order simply to eliminate supply sensitivity.

The concept of *national security* also needs to be briefly considered before we can proceed with our analysis. In this matter I fully agree with Richard H. Ullman's view that decision-makers and scholars alike tend to conceive of it in excessively narrow and essentially military-strategic terms.[7] A major reason for this common conception is that it is considerably easier, both politically and intellectually, to posit national security in this manner rather than with reference to seemingly less tangible perils or values. The consequence of such a view is, however, that it tends to ignore not only other—and perhaps equally or even more insidious—threats to a nation but also helps to reinforce rather than ameliorate the militarization of international conduct. To regard the world exclusively in the classical realist terms of high politics is surely to close one's eyes to the equally pervasive threats to a nation's well-being which are a function neither of foreign armies nor military hardware but of economic, technological and similar forces of international domination and power. Indeed, as Barry Buzan has noted, 'The idea of the state, its institutions, and even its territory can all be threatened as much by manipulation of ideas as by the wielding of military power'.[8]

For these and similar reasons, the present discussion will regard 'national security' in the less conventional way proposed by Ullman. He suggests that we view a challenge to national security as

> an action or sequence of events that (1) threatens drastically and over a relatively brief span of time to degrade the quality of life for the inhabitants of a state, or (2) threatens significantly to narrow the range of policy choices available to the government of a state or to private, nongovernmental entities . . . within the state.[9]

Similarly, Klaus Knorr has noted that in so far as national security queries inevitably arise when vital national values are perceived to be threatened, 'there is no reason why economic values and particularly patterns of economic life cannot be regarded as vital'.[10] Clearly one such threat to vital values are interruptions in the flow of critically needed natural resources, be they a function of the international expansion of demand (i.e. market forces) or of restrictions on supplies (intended or not). A signal benefit of this conception of national security is that it allows us to postulate the general premise—proposed by Ullman—that in all spheres of policy, security increases as vulnerability decreases.[11] Although formidably difficult to measure, this axiom will constitute a cornerstone of the present analysis.

Since *oil disruptions* are the immediate cause of the types of crises sketched above, any discussion of oil vulnerability must start with an examination of this specific issue. At the same time this factor must bs distinguished from two other dimensions of the energy supply issue as a whole. The first of these refers to the multiple problems associated with the high level of imports which—as we have noted above—characterizes most oil-importing countries and has done so for many years. *Dependence on oil imports* is essentially an issue which can only, at best, be tackled over a time-span of decades, while (it will be argued below) vulnerability is in need of, as well as in practice amenable to, policies which relate to the immediate future. Both of these problems should also be distinguished from the long-term *depletion of world oil resources*. Since the resolution of this issue requires the development of alternative energy technologies with extensive lead times, its impact can only be felt over a much longer time period.[12]

Although these three aspects of the energy problem are obviously closely intertwined, the present discussion will refer mainly to the first issue. In other words, we will here be addressing the short-term problem associated with disruptions of oil supply which can occur at any time, and which because of this also requires immediate solutions. In Robert J. Lieber's terminology, we will here focus on energy 'crunches' rather than on energy 'squeezes'.[13]

In view of these sets of concepts, what have been the basic structural characteristics of the past oil crises? In a most skeletal fashion, energy vulnerability as here conceived can be related to two major mechanisms which tend to cause supply disruptions to threaten the economies—and hence the national security—of oil-importing countries in the West. The first of these is connected with dramatic price increases, which lead policy-makers to attempt the substitution of alternative fuels and other economic resources for oil. In so far as the scope for such substitution and adaptation is limited in the short run, the result is a drop in the gross national product and an increase in unemployment. Since it has been estimated that a 10 million barrels a day (mbd) disruption of Persian Gulf oil supply during one year can cause a 4 to 8 per cent GNP loss for the United States, and a 6 to 11 per cent loss for Europe and Japan, such disruption losses can obviously become quite significant.[14]

The second manner in which such disruptions affect economic losses is more indirect but no less tangible in its effects. This mechanism comes into play when governments try to counteract rising oil prices by imposing price controls and oil allocation schemes on their domestic fuel markets. Such controls can add significantly to the level of disruption losses by giving the energy market wrong price signals, leading to physical shortages and large indirect losses.[15] Obviously, the separate and combined effects of these mechanisms will depend both on the type and the level of actions taken during disruptions.

Although it is often assumed that oil disruptions from the Persian Gulf are necessarily the consequence of purposive OPEC behaviour, this is clearly not the case, as the two most recent disruptions have illustrated so well. On the contrary, even a cursory examination of the geographical, demographic, political and military situation of this region will indicate why the flow of oil from it is constantly vulnerable to a whole series of political perturbations, military adventures or other 'accidents' and mishaps. This is, after all, a region which over the last three decades 'has been subjected to a half-dozen wars, a dozen revolutions and innumerable assassinations and territorial disputes', as Robert Stobaugh and Daniel Yergin have laconically noted.[16]

The high disruption risks of this area can, in a most unsystematic fashion, be classified in terms of at least the following factors: (1) the instability of the so-called Northern Tier, which has been so abundantly exemplified by the events of 1978 onwards; (2) internal political instabilities in Saudi Arabia, the Emirates and Qatar (including acts of sabotage); (3) the downfall of the present Saudi regime, followed by those of the Emirates and Qatar; (4) various forms of new embargoes in connection with the Arab–Israeli conflict; (5) an outlet disruption of the Strait of Hormuz by either Iran or Iraq (in connection, perhaps, with Oman); (6) resource-management strategies on the part of the major producers aimed at slowing down the rapid depletion of existing oil reserves; (7) shipping disruptions due to superpower conflicts or interventions; and (8) technical accidents to major oil-producing facilities.[17]

These potential causes or catalysts of disruptions should, furthermore, be linked to a more general condition which characterizes the contemporary international oil market, making it much more unpredictable and complicated than the situation typical of the first postwar decades. Indeed, it has been maintained that few sectors of the international economy have experienced such a profound change in such a short period—roughly, between 1970 and 1980—as the international oil market.[18] Basically, this change is a consequence of the nationalization of oil extraction in most of the oil-producing countries, followed by the increased nationalization of oil trade as well. These changes are not merely the result of economic or market fluctuations but part of an historical process which seems to bear most of the hallmarks of permanency.[19] It involves a change in the bargaining relationship between oil exporters and importers, a process in which oil itself has become less and less a conventional commercial commodity—such as pork bellies or copper—and increasingly a political commodity in an arena of international instability. This politicization of the oil trade, with governments playing central roles at both ends of the market in so-called government-to-government (or G–G) oil deals, has provided a high degree of selective choice to the suppliers of oil, *inter alia* activating oil supplies as a political instrument—the so-called 'oil weapon'—for pursuing other than strictly commercial interests.[20] The result has been the emergence of patterns of 'politics through trade' rather than profit maximization. Additionally, since oil-importing governments have been willing to give political favours in return for preferential treatment by the oil-producing states, there has also emerged a concomitant pattern of 'trade through politics'.[21] In so far as politics itself is such an inherently volatile factor in the Persian Gulf, these new trade premises, often linking the sale of oil to non-commercial issues, have exacerbated an already very unsettled market situation. Ernest J. Wilson has in this connection thus referred to politicization in terms of 'fundamental challenges to the rules and regulations of the market', resulting in a far-reaching impact on both world economics and international politics.[22] Indeed, this development has in some quarters even been linked with Gresham's Law, the implication being that during these crisis years bad politics managed to drive out good economics.[23]

Assessing the potential occurrence of any of these disruptive factors is, of course, a task which must be approached with considerable circumspection. At the present time Northern Tier instabilities probably remain potentially the most explosive.[24] On the other hand, although his is not a view which is generally accepted by all oil experts, Walter Levy's analysis points to the conclusion that the principal threat to a stable flow of oil during the remainder of this century may stem from developments within individual OPEC countries rather than from conflicts between them or between the oil-importing states and OPEC.[25] In any event, it is probably correct to suppose—as Wilfred Kohl has done—that the overthrow of the present Saudi regime would be perceived in the West as the ultimate nightmare.[26] It is also realistic to assume, as Lieber does, that the odds that some kind of crisis-precipitating event will occur are overwhelming.[27] Although this assumption was made in the early 1980s, nothing has really happened since then to make it out-of-date.

AVAILABLE POLICY CHOICES

Given the ever-present threat of oil disruptions, it is imperative to make a realistic assessment of the remedial responses which in principle are available for meeting it. Otherwise, as David A. Deese and Linda B. Miller have noted in a major Harvard study, oil-importing nations may simply find themselves preparing, 'like army generals, to refight the last war'.[28] Although there are obviously no quick fixes or one-shot solutions to this problem, the recent literature on it nevertheless provides us with an increasingly consensual basis for reaching such a realistic appraisal.

There is, first of all, a very strong agreement on the utter necessity of maintaining sufficient *emergency stockpiles*. Since the oil crises of the past years have time and again brought home the lesson of the inelasticity of oil demand, leading even small interruptions to drive up prices enormously, it has become obvious that the governments of oil-importing nations are dependent on the availability of strategic stockpiles if they are to be able to dampen this effect and its consequences. It has also been demonstrated that the major benefit of a stockpile release comes from its effect on unemployment and the deleterious consequences following in the wake of the latter. In other words, the appropriate release of stockpiled oil has the effect of counteracting the first economic mechanism which was referred to above.

James Plummer and his Stanford associates have provided us with an extensive study of this effect in terms of their discussion of the payoff premium of oil stockpile policies. After a painstakingly thorough analysis of this premium in terms of 'price-reaction functions' and 'buying-power wedges', they conclude that the best measure of oil disruption preparedness is the ratio of non-working oil inventories to oil consumption.[29] The basic idea is that by substituting for imports during interruptions, the release of stockpiles can reduce rising world oil prices and hence the economic effects of such interruptions. Given a tight market and the low price elasticity of oil demand, even small supply disruptions can have dramatic consequences on the spot market (both in terms of volumes traded and prices paid), which in turn affects price run-ups generally. Lieber has in this connection surmised that 'had the consuming countries built up stocks by 1.2 mbd in the non-crisis year of 1978 and then drawn them down at the rate of 0.8 mbd throughout 1979—instead of doing precisely the reverse—much of the price shock would have been avoided'.[30] He has also noted that whereas competitive spot market bidding causes steeply rising prices for oil, such behaviour does not in itself add to short-term supply availability.[31]

Of course, buying extra oil to add to a stockpile is expensive, not least because of the incremental costs which are involved when increased demand drives up prices even during 'normal' years.[32] However, stockpiling is a form of insurance, and insurance premiums always seem expensive during normal times. Its value can only be fully appreciated during emergencies, when an oil stockpile can help to avoid both huge losses in GNP and their concomitant economic and social effects. Undoubtedly, as Hanns Maull has noted in this connection, one problem is that risk-insurance in this area is complicated by the fact that while 'worst-case insurance would probably involve very massive costs . . . a low-insurance policy might be insufficient to avoid major dislocations'.[33] But this lack of appropriate actuarial tables can never be construed to constitute an argument against the utter necessity of realistic stockpile policies—the risks involved are simply too high. As Edward N. Krapels has noted in his extensive analysis of emergency stockpiles, while such inventories are a ' "dead asset" of considerable proportions', they are nevertheless 'an essential element of international oil security, in the same way that armaments . . . are essential to military security'.[34] Indeed, in this perspective armaments are equally 'dead' as assets; yet we take our huge military budgets and stockpiles for granted, while balking at carrying sufficient insurance against oil supply interruptions.

A second method for ameliorating the effects of supply disruptions involves the application of international *disruption allocation programmes* among the oil-importing nations. Following the Yom Kippur War, this approach was made into a cornerstone of the International Energy Programme (IEP) and hence of the IEA, both established by the agreement signed in November 1974.[35] Indeed, is so far as the primary objective of the IEA is to limit damage from oil supply disruptions by means of an equitable allocation of the remaining supplies between member countries (although reducing the total level of import demand is a related goal), most of the operational—and highly complicated—mechanisms incorporated in this agreement deal exclusively with this instrument.

Basically, these mechanisms incorporate stipulations regarding the maintenance of emergency reserves, and triggering provisions for (1) carrying out restraint measures when a given level of supply reduction is either anticipated or sustained; and (2) allocating supply rights to the participating members of the programme.[36] Since the IEP relates mainly to the supply problem (reflecting its genesis in the embargo atmosphere of early 1974), its provisions with respect to the price of the oil thus allocated is deliberately ambiguous, referring merely to the 'price conditions prevailing for comparable commercial transactions', as the agreement notes.[37] In effect, this means that the IEA is not empowered to deal explicitly with the price problem during supply interruptions.

The consequences for any given participating country of an activation of the programme will depend heavily on its ability to carry out restraint measures with a minimum of economic pain. This ability is complicated by the variety of uses to which energy can be put, and by the means available to substitute one form of energy for another. No two countries are alike in these respects and hence to how a curtailment of oil supplies will affect their respective economic sectors. Thus, for example, while in 1973 oil was used in France to fuel almost 60 per cent of its power plants, the comparable figure for Germany was less than 10 per cent.[38] In view of this factor, it is obvious that equivalent disruptions of oil imports will have substantially more deleterious effects on the French than on the German type of economic structure.

Krapels has given the following reason why differences such as these were not taken into consideration when the programme was first designed:

> the IEP allocation program that emerged in 1974 . . . is a political compromise. . . . [It] did not take into account differences in the efficiency of national energy utilization or in the degree of oil dependence. It did not take into account differences in the extent of dependence on oil imports.[39]

Such fundamental domestic differences—and they are many and complex—pose severe problems for the governments participating in the programme, at the same time as they stress the need to prepare individually for the collective burdens which disruption allocation schemes of this kind require.

Policies to *reduce oil imports* prior to anticipated disruptions constitute a third important avenue for reducing energy vulnerability. This option involves a choice between two basic policy instruments, namely conservation and the substitution of other forms of energy—preferably domestic—for imported oil.[40] The essential aim of these policies is to lower the base level of imports at the start of an interruption and, by extension, to make a country economically less dependent on the supply of imported oil.

To analyze its potential effects, Plummer and his colleagues have posited the concept of the 'oil import reduction premium', defined as the dollar economic benefit resulting from reducing oil imports by one barrel. Here again they have used the notions of the buying-power wedge and price-reaction functions to measure it; and they have found that the benefits of this instrument arise from two different sources. First, lowering imports during normal years has the effect of reducing world oil prices and hence the oil import bill of a country. (Here, contrary to the case with the oil stockpile premium, the buying-power wedge is positive, for the simple reason that reducing demand reduces the price of oil, while increasing demand—for instance, when filling empty inventories—drives up the price.) Second, having lower import needs during the onset of an interruption reduces interruption losses, since the resulting oil price increases are paid on lower volumes of oil. Together, these two effects tend to reduce macroeconomic losses during interruptions, though not by as much as in the case of stockpile measures.[41]

A problematic aspect of the oil import reduction premium is that it is difficult to compare it with the stockpile premium. As noted by the Stanford team, there are obviously 'trade-off and synergisms between oil stockpiling and oil import reduction'.[42] More fundamentally, there is also a contradiction between a policy of buying more and buying less, except if stockpiling can be accomplished without increasing imports. For an oil-producing country such as the United States this is of course entirely feasible; but this does not apply to the majority of IEA's less fortunate members.

Plummer and his associates conclude that while both measures are urgent national objectives, their analysis tends to favour a policy of stockpiling over oil import reduction policies.[43] Lieber is perhaps somewhat more equivocal when he enjoins that

> specific measures to deal with the energy problem must be weighed against whether they meet one or both of two key criteria. Either they must reduce dependence on imported oil, and thereby respond to basic causes, or they must provide the means to lessen the disruptions which will be caused by future oil crises.[44]

As short-term instruments, both stockpiling and—if to a lesser extent, since we are here concerned with the effect of price run-ups in the event of interruptions—allocation programmes are probably more effective, even though they do not touch the basic causes of the problem. To the extent that a policy of reducing oil imports deals more directly with these causes, it is obviously to be preferred—as a middle- to long-range measure.

Aside from the individual merits and problems associated with each of these measures, it must be stressed that their individual and joint effectiveness will to a large extent depend on extensive international co-operation between the oil-importing states. In this connection three different situations are pertinent; and each, as Lieber notes in his elaboration of them, requires a different mixture of measures to counter the effects of oil vulnerability.[45]

The first type of situation is not only the most normal but also the category which offers the best opportunities for extensive co-operation. I here have in mind non-emergency periods, such as those which occurred between 1975 and 1978, and after 1981, during which policies requiring long lead periods and extensive planning can be co-ordinated. Lieber has posited the following types of measures as the appropriate focus for co-operation during these non-crisis situations: (1) adaptation of international energy patterns, with the purpose of diversifying sources of imported oil, increasing the production of non-OPEC as well as other energy sources and improving energy efficiency; (2) adoption of economic and financial measures aimed at easing the negative effects on the oil-importing countries of the transfer of wealth to the producing countries; (3) preparation for crisis management by agreeing on emergency measures, increasing stocks and providing guidelines for their use; (4) diplomatic co-ordination of relations towards the oil-producing countries; and (5) military contingency planning vis-à-vis the Persian Gulf, including efforts to deter threats of energy supply in this region.

The second type of situation occurs when small-scale supply interruptions become a reality, as they did in 1973–4 and once again in 1979–80. During such periods the aim of co-operation is to restrain price increases, to avoid a competitive scramble and to co-ordinate the strategies of the oil-importing countries. Specific measures include the following: (1) co-ordinated draw-down of petroleum stocks; (2) agreements aimed at decreasing demand pressures; (3) the redistribution of petroleum; and (4) measures aimed at alleviating the price shocks of spot market bidding.

The third and most serious case occurs when the general petroleum shortfall during a crisis passes beyond the 7 per cent mark—the trigger level for IEA's emergency sharing system. This is, as Lieber notes, the least likely category of risk but also potentially the most dangerous situation in which the oil-importing countries can find themselves. Despite the October War, the Iranian revolution and the continuing war between Iran and Iraq, overall international—as distinguished from national—shortfalls of this magnitude have as yet not occurred. However, if this situation were to materialize, the following co-operative measures would be applicable: (1) emergency oil sharing under the aegis of the IEA; (2) political co-ordination of energy policies, aimed at avoiding open disagreement; and (3) participation or acquiescence in military measures.

POLITICAL RESPONSES

It would be an understatement to claim that the actual responses to the supply interruptions of the 1970s were inadequate. Why was this the case?

Clearly, the answer to this query is no less complex than the—seemingly innocuous—query itself. Even the apparently easy task of assigning blame (or finding scapegoats) is in fact neither simple nor uncontroversial. As Plummer has noted in this connection, there 'is plenty of blame to spread among all groups—consumer advocates and their political allies, international political analysts, conservationalists and environmentalists, engineers and scientists, and economists'.[46] Although this approach as a genre has handsomely sustained a whole literature over the past decade, I shall here avoid the dubious pleasures of such an inquiry and instead briefly outline the actual reactions of the oil-importing community to the crises of the past decade.[47] This will, it is hoped, facilitate a dispassionate discussion of the problem at the same time as it may suggest the analytical and political wherewithal for avoiding the mistakes of the past. More importantly, such a discussion will provide us with a general background for subsequently discussing the specific Swedish case. For the sake of brevity and analytic convenience, this inquiry will consist of a brief consideration of the extent to which—and to what effect—each of the instruments discussed above has been utilized to ameliorate the economic effects of sudden oil disruptions. Since we shall deal only with the common rather than the individual responses of the countries in question, this evaluation will of necessity take the form of somewhat superficial generalizations.

With respect to stockpile policies, the recent literature reflects an almost unanimous agreement that these have been far from adequate to counteract the economic effects of disruption losses. Clearly this was the case in 1973–4, which is perhaps understandable; but as we have already noted, the same conclusion also applies to the second crisis, despite the loose oil market of the intervening years and the introduction, if not full application, of IEA provisions with regard to emergency stockpiling. Furthermore, this state of affairs has persisted to the present time, despite American and Japanese policies aimed at acquiring emergency reserves more in line with the IEA agreement.

More fundamentally, however, it has also been shown that even adhering to the IEA norm—of keeping a stockpile equalling at least ninety days of net imports—does not by itself constitute a guarantee against the adverse economic effects of sudden interruptions in the flow of oil. In Krapels's view, 'many European governments seem to have based their emergency reserve size on an optimistic view of the probability, level, and duration of supply emergencies'.[48] As against this optimism, Plummer has argued not only that the IEA norm is too weak because its definitions of stockpiles include many so-called working inventories (which hence do not constitute additionally usable oil during disruptions) but also that much higher levels of 'true' stockpiles are needed to achieve adequate protection against the energy vulnerability of the oil-importing countries. He also points to IEA's weakness of not possessing mechanisms for co-ordinating stockpile drawdowns during a disruption, or for integrating stockpile drawdowns with the rest of the emergency sharing system.[49] With respect to the United States, these factors have led him and his associates to advocate a strategic stockpile of at least 2 billion barrels, which is approximately eight times as much as the existing stockpile was in the early 1980s (when this recommendation was made). Since the United States at this time imported about 8 mbd, this minimum recommendation amounts to keeping a stockpile equivalent to 250 days of net imports. Although not as thorough and specific in their evaluations of the critical importance and present inadequacy of emergency inventories, similar recommendations can be found in almost all of the more recent studies of energy vulnerability.[50]

It is also generally acknowledged that the reaction of the oil-importing countries to the 1973–4 oil embargo was far from satisfactory with respect to providing a co-ordinated programme for allocating oil to those worst hit by it. Deese and Miller thus note that 'Pursuit of national interests and unwillingness to share supplies characterized European responses to the 1973 embargo'.[51] Nor did matters improve once the crisis was over. In the words of the Stanford study:

> One would think that the suddenness of these events and their large economic consequences would have prompted the oil importing countries to take strong policy actions to reduce their vulnerability to further oil supply disruptions and price increases. That did not occur. Between 1974 and 1978, little happened other than a torrent of words uttered in debates over energy policy.[52]

Instead, once the immediate crisis had subsided and supplies started flowing downstream again, complacency once more became the normal response of the general public and most decision-makers.

The reactions to the second crisis have similarly demonstrated how ill-prepared and essentially unwilling or unable the oil-importing countries were to co-ordinate—rather than simply react to—oil supply emergencies. Lieber, reporting on a private résumé of a leading IEA official, thus writes that

> the experience of 1979 was a bad one: All agreed to do something, but no one did anything. Action continued to be taken on a *suave qui peut* basis; although this was veiled by nice-sounding communiqués issued at high-level meetings, there was no *real* effort to face the situation as an international community.[53]

Instead, a Harvard study reports, 'bilateralism in the form of government-to-government deals has become as permanent fixture of the oil marketing and consumption pattern for many European states'.[54] More recently, Wilson has expanded on the significance of this factor as follows:

> the catastrophic 1978–80 price increase (160%) was caused not by physical shortages of oil but, rather, by the rise of state-to-state oil transactions as a major new form of industrial organization. There was easily enough oil in the world; what was missing were readily available distribution channels through which to move the oil. The old oil

channels—the multinationals—had been squeezed as a result of dramatic changes, and the new channels—the G–G oil deals—could not, or would not, handle the suddenly increased pressure.[55]

Krapels draws the obvious conclusion from these developments when he notes that 'If the relatively minor disruption of 1979 could not be managed effectively, the prospects of managing a more severe crisis appear dim'.[56] And since the small price increases following upon the 1980 events were due mainly to the unplanned—if fortuitous—conditions created by the disruption immediately preceding it, it too affords us little comfort.

Various reasons can be adduced to explain the frailty of the IEA disruption allocation programme. First of all, since IEA mechanisms for invoking emergency oil-sharing only come into effect above shortfalls of 7 per cent, they have proved to be irrelevant during the type of crisis which was produced by the Iranian revolution (when only a temporary shortfall of 4 per cent occurred). Second, the industrialized countries belonging to the IEA obviously have fundamentally different interests with respect to energy issues. Some are both producers and consumers of oil, while others are wholly dependent on the import of crude for their petroleum needs. This factor—and the ensuing anxieties on the part of most and the increasing export competition between all (to pay for oil imports and to reduce balance of payment problems which such increasingly expensive imports create)—is obviously not conducive to a spirit of co-operation.[57] Also, given the cynical view expressed by some knowledgeable observers that the IEA was in the first place created to protect its members from one another (or, more likely, given its genesis in Henry Kissinger's Washington, to pursue American interests), it should not surprise us that there exists such a widespread scepticism about its ability to foster unified action when faced with a crisis.[58]

Third, the fact that the IEA allocation procedures have as yet not been allowed to show their mettle in reality has also led many observers to question the probability of their success when or if faced with a real contingency.[59] This judgement applies in particular to a situation forcing the IEA countries to face an open-ended crisis. As Krapels has put it:

> A disruption caused by deliberate supply restrictions for political purposes would threaten the cohesion of the IEA on several fronts: it would raise the ugly prospect of emergency reserve depletion; provide enough time for the difference in importing countries' vulnerability to become apparent; and being inherently political, cause 'hostage' members to resent the actions of the 'targets' that brought on the disruption.[60]

Fourth, the feasibility of joint action programmes such as the allocation scheme presupposes the existence of effective individual mitigation measures; otherwise—and this is one of the greatest weaknesses of the IEP as a whole—some participating countries may in the event of a real crunch opt to go it alone rather than sharing a burden which is not perceived to be either economically equitable or politically feasible. It is certainly not obvious that the required mitigation measures have been equally effectively pursued by the member countries of the IEA. As it has been noted time and again, public disbelief in the existence of a real energy problem was quickly in evidence after each crisis, making it politically and otherwise problematic to pursue coherent policies requiring expensive pre-emptive measures.[61]

A final consideration pertains to the future relationship between the United States—not only the world's largest oil importer and supplier of energy technology but also one of its largest producers of oil—and the IEA. In this connection Louis Turner has suggested that the 'IEA . . . is a long way from Washington and an Administration like the Reagan one which believes that market forces can solve most problems is inevitably not over-enamored with an Agency concerned with emergency oil allocation and energy RD&D schemes'. Since, in his view, it is very hard to see how the energy allocation scheme could survive without the United States, and since the Reagan administration is refusing to renew its authority to allocate oil supplies during a crisis because it feels that adequate provision already exist, we here clearly have a potentially grave danger to the whole IEA allocation system itself.[62] As a whole, these sceptical views on the workability of the scheme provide, if anything, an additional argument for the wisdom of substantially increasing national stockpiles well above existing IEA requirements.

The picture with regard to the reduction of oil imports is somewhat less gloomy. The long-term trend toward increased dependence on oil was in fact reduced after the first energy crisis; and this pattern has been accentuated during the most recent years. Although oil consumption remained flat during 1979, it declined by about 7 to 8 per cent in 1980 and an additional 5 to 6 per cent in 1981.[63] This downward trend has continued until the very recent past.[64] At the same time non-oil energy consumption increased by 5 per cent in 1979, and by 2 per cent each of the following two years. Both nuclear energy and, in particular, coal have been substituted for oil. Thus, for example, Japanese coal imports increased by a quarter between 1977 and 1980, while international trade in coal experienced a 30 per cent boost between 1978 and 1981. More importantly, the most substantial adjustment has occurred on the demand side in the form of a greater energy efficiency. Thus, while the total primary energy consumption in the industrialized world between 1968 and 1973 grew annually at a rate of 5.6 per cent, the comparable figure for the period between the two energy crises was an average of only 1.2 per cent.[65]

Of course, a major reason for this drop in demand is to be found in the reduction of economic growth characterizing these years, which in turn can be linked to the various oil supply disruptions and the dramatic price increases which they caused.[66] Clearly, with the onset of a severe recession in 1974, leading to postwar records with respect to drops in GNP and increases in both unemployment and inflation (together constituting, as previously noted, the 'stagflation' phenomenon so characteristic of the years in question here), the demand for ever dearer oil was bound to drop drastically.

Nevertheless, although GNP and energy consumption during the 1960s grew at the same rate, a partial decoupling of this link has occurred since then. As Yergin has noted, the divergence between these two measures is such

> that by 1981, the industrial world consumed about 5 per cent more energy than it had in 1973, despite a real GNP growth of about 21 per cent. In other words, the industrial world used 16 per cent less energy in 1981 than would have been the case if the previous one-to-one relationship had been maintained.[67]

This broad trend toward energy efficiency (coupled with the severe economic downturn) also had a dramatic effect on the oil market of the early 1980s. Demand for OPEC oil thus fell no less than 40 per cent over a three-year period—from 31 mbd in 1979 to less than 18 mbd in early 1982. Unfortunately, this development has also had the effect, once more, of creating a glut market and with it a glut mentality: the belief, increasingly widespread again, that the energy problem is behind us for good, leading to growing public and political resistance to contingency and conservation measures. This loss of momentum is especially worrisome in view of the revised policies of the present American administration, which clearly tend—and seem intended—to reinforce the assumptions underlying this glut mentality. President Reagan, a month after his first election to office, thus jokingly mused about conservation as a policy which simply means that 'we'll be too hot in the summer and too cold in the winter'.[68] Since the role of the United States is as crucial in reducing oil imports as it is in the collective allocation of petroleum resources during disruptions, this type of response and the ensuing policy of '*laissez-faire* liberalism' is very regrettable indeed—especially if, as Yergin suggests, a sudden oil price rise in the future will most probably have an even greater negative effect on oil-consuming countries than the first or second crisis did.[69]

Since about half of the oil moving in world trade flows through the Strait of Hormuz, protecting it and the Persian Gulf physically from military and similar perturbations remains a constant priority. American perceptions of this situation are pre-eminent—indeed, more so than with respect to the other measures discussed above—at the same time as the military-political situation itself is anything but simple, since a viable military policy must cover a whole spectrum of options, ranging (as we have noted earlier) from local insurgencies to a full-fledged Soviet military invasion of the area. As the Harvard study above has suggested, the ability of the United States (and her allies) to meet these objectives is a function of three interrelated factors: the composition, strength and availability of Western forces for deployment to the Gulf; the military and political responses of local governments; and military threats to Western interests elsewhere in the world. 'In this regard', Geoffrey Kemp adds, 'there are significant constraints upon current US capabilities that could probably not be overcome quickly, even if top priority and greater funds were assigned to Gulf contingencies'.[70] Due to current divisiveness in Western policies, a general agreement on a plan of sufficient scope remains elusive, at the same time as Soviet opportunities for interference have by no means decreased.[71] In their parallel Stanford study, Henry S. Rowen and John P. Weyant claim that although 'the actions we and our allies have taken in the past year have been valuable, they have not been enough to reverse the balance' in favour of the Soviet Union's superior capacity to project into the Gulf.[72] They conclude that relative to the magnitude of the losses which a closure of the strait would cause, 'it is obvious that there have been a woeful underinvestment in . . . military preparations' on the part of the United States and her allies.[73]

However, it must be added that at least on this score the Reagan administration—true to its policy of greatly enhancing American military capabilities in general—has acted decisively to reverse this type of 'underinvestment'. Thus it has been noted recently that 'The declaration . . . early in 1984 that the United States would keep the Strait of Hormuz open in the event of an Iranian blockade illustrates the increased military role which the United States saw for itself in the area towards the end of the period'.[74] Today, of course, this policy reversal has taken on an explosive reality all of its own.

CONCLUSION

As recently noted in a review of the first two oil shocks, 'Energy policy must be anticipatory, not reactive'.[75] Clearly the former type of policies was not only absent prior to the first crisis, which is understandable, but also, on the whole, prior to the second, despite the lessons which the 1973–4 experiences should have taught decision-makers throughout the industrial world. It is obvious, however, that different nations acted differently in this respect, and that this is reflected in how much political energy and financial resources were spent on increasing the anticipatory level of their respective energy policies during the 1970s.

In the following chapter we shall take a closer look at Sweden's record in this respect. The aim of this chapter is, however, not only to gauge the anticipatory level of her energy policies but also to pave the way for a discussion of these in relation to Sweden's particular brand of national security policies. The question towards which this analysis is intended to move is the following: were Swedish energy policies during the decade in question commensurable with the basic assumptions of the logic underlying her very special type of foreign policy doctrine?

NOTES

1. See David A. Baldwin, 'Interdependence and Power: A Conceptual Analysis', *International Organization*, vol. 34, no. 4 (Autumn 1980), pp. 481ff.
2. Robert O. Keohane and Joseph S. Nye, Jr, *Power and Interdependence* (Boston, 1977).
3. Ibid., pp. 491f.
4. This conceptualization of 'dependency'—but not of 'vulnerability'—thus differs from the one which Nye has suggested when he writes that while oil 'dependency is measured by the level of imports . . ., degrees of vulnerability are judged not by the level of imports, but by the costs of interruptions *after* considering the effectiveness and costs of constructing measures to escape the situation'. Joseph S. Nye, Jr, 'Energy and Security in the 1980s', *World Politics*, vol. 35, no. 1 (October 1982), p. 128.
5. See Kenneth N. Waltz, 'The Myth of National Interdependence', in Charles Kindleberger (ed.), *The International Corporation* (Cambridge, Mass., 1970), passim, and Waltz, *Theory of International Politics* (Reading, Mass., 1979), pp. 139ff.
6. Waltz, *Theory of International Politics*, p. 142.
7. Richard H. Ullman, 'Redefining Security', *International Security*, vol. 8, no. 1 (Summer 1983), pp. 129ff. In this connection, see also Kal Holsti's recent definition of 'power': the ability to 'protect autonomy, maximize latitude of choice in policy making, reduce constraints emanating from abroad, and maintain domestic political legitimacy and authority'. He is surely correct to conclude that 'Power in this sense is really synonymous with the concept of security', and that it does not as such 'signify only military capabilities, territorial expanse, or prestige'. Kal J. Holsti, 'Politics in Command: Foreign Trade as National Security Policy', *International Organization*, vol. 40, no. 3 (Summer 1986), pp. 645f.
8. Barry Buzan, *People, States and Fear: The National Security Problem in International Relations* (Brighton, 1985), p. 65.
9. Ibid., p. 133.
10. Klaus Knorr, 'Economic Interdependence and National Security', in Klaus Knorr and Frank M. Tragle (eds.), *Economic Issues and National Security* (Lawrence, Kansas, 1977), p. 8.
11. Ullman, p. 146. Obviously, this axiom assumes a *ceterus paribus* clause.
12. James Plummer, 'Energy Vulnerability Policy: Making It Through the Next Few Years Without Enormous Losses', in James Plummer (ed.), *Energy Vulnerability* (Cambridge, Mass., 1982), pp. 5f.
13. See Robert J. Lieber, *The Oil Decade: Conflict and Cooperation in the West* (New York, 1983), pp. 120ff.
14. See Plummer, pp. 2f, and Joseph S. Nye, 'Energy and Security', in David A. Deese and Joseph S. Nye, Jr, (eds.), *Energy and Security* (Cambridge, Mass., 1981), pp. 3f.
15. Plummer, p. 2.
16. Robert Stobaugh and Daniel Yergin, 'The End of Easy Oil', in Robert Stobaugh and Daniel Yergin (eds.), *Energy Future* (New York, 1980), p. 5.
17. See James Plummer, Hung-Po Chao, Dermot Gately and Richard Gilbert, 'Basic Concepts, Assumptions, and Numerical Results', in Plummer, pp. 15ff; Wilfred

L. Kohl, 'Introduction: The Second Oil Crisis and the Western Energy Problem', in Wilfred L. Kohl (ed.), *After the Second Oil Crisis: Energy Policies in Europe, America, and Japan* (Lexington, 1982), p. 4; Gordon T. Goodman, Lars A. Kristoferson and Jack M. Hollander (eds.), *The European Transition from Oil* (London, 1981), pp. viiff; Nye, 'Energy and Security', pp. 3f; Nye, 'Energy and Security in the 1980s', pp. 132ff; Edward N. Krapels, *Oil Crisis Management* (Baltimore, 1980), pp. 20f; Robert J. Lieber, 'Middle East Oil and the Industrial Democracies', Working Paper No. 22, International Security Studies Program, Woodrow Wilson Center for Scholars, March 1981, pp. 1ff; and *Newsweek*, 13 February 1983, pp. 8ff.

18. See in particular Oystein Noreng, 'State Oil Trading and the Perspective of Shortage', in Paul Tempest (ed.), *International Energy Options* (Cambridge, Mass., 1981), pp. 229ff, and Paul Kemezis and Ernest J. Wilson III, *The Decade of Energy Policy: Policy Analysis in Oil-Importing Countries* (New York, 1984), pp. 3ff.
19. See Noreng, p. 229. It should here be added, however, as Ernest J. Wilson III has recently emphasized, that 'the oil market has always had a considerable admixture of the economic and the political. The oil industry is *strategic* for military and commercial purposes; it constitutes a large industry in a domestic economy and the largest single component of world trade in dollar terms; many nations' companies compete for prominence within the industry; it is subject to periodic cyclical instability that often calls for government concern and action'. As noted by Anthony Sampson, Louis Turner and others, 'the oil industry had always been deemed too important to be left to private companies, just as war cannot be left solely to generals'. Ernest J. Wilson III, 'World Politics and International Energy Markets', *International Organization*, vol. 41, no. 1 (Winter 1987), p. 143.
20. Plummer et al., p. 15. According to Wilson, however, the commercial and political claims made for these G–G deals have been exaggerated by both consumers and producers. '"Political" considerations', he writes, 'will play a larger or smaller part in corporate decision-making as the world environment changes, but over time they can never totally replace commercial criteria'. Wilson, p. 143. For a discussion of five different notions of market politicization, see the same author's 'US International Energy Policy: Competing Explanations of Market-Politics Interactions', paper presented to the European Consortium for Political Research (ECPR) Joint Sessions of Workshops, Salzburg, April 1984, pp. 18ff.
21. Noreng, p. 238.
22. Wilson, 'US International Energy Policy', pp. 19f.
23. Ibid., p. 20.
24. See Plummer et al., p. 15.
25. Walter Levy, 'Oil Policy and OPEC Development', *Foreign Affairs*, vol. 57 (Winter 1978–9), pp. 298ff. See also Wilson, 'World Politics', pp. 144ff.
26. Kohl, p. 4.
27. Lieber, 'Middle East Oil', p. 12.
28. David A. Deese and Linda B. Miller, 'Western Europe', in Deese and Nye, p. 183.
29. James Plummer, 'Conclusions: Analytical Results, Directions for Policy, and Important Unanswered Questions', in Plummer, p. 344. For their discussion of these concepts, see pp. 22ff. The *stockpiling premium* is defined as 'the expected

dollar economic benefit, in terms of lower losses in disruption years, of adding a barrel of oil to stockpiles during normal years. This premium is a net measure that takes account of the fact that filling the stockpile during normal years will cause somewhat higher world prices in those years' (p. 27). The *price-reaction function* is basically a 'curve illustrating the net supply available to the importer at various prices' (p. 24). The *buying-power wedge* refers to the 'divergence between the private cost of imports, as represented by the world oil price, and the higher social cost of imports' (p. 21). In other words, it represents the difference, in price, between the price-reaction function and the total cost of imports. Hence, the 'total value of stockpiled oil in avoiding disruption losses is the world price (in the non-disruption year when stockpile filling occurs)' plus the buying-power wedge plus the stockpile premium. 'The marginal social cost of acquiring a barrel of oil for the stockpile is the world oil price plus the buying-power wedge. . . . The residual between these two magnitudes is the oil stockpile premium. . . . Once the stockpile premium does get down to the cost of stockpiling, then the optimal stockpile fill rate is zero, and the equilibrium stockpile size has been reached' (pp. 31f).

30. Robert J. Lieber, 'Energy and Foreign Policy: Western Responses to the Oil Crises', paper presented to the XII World Congress of the International Political Science Association, Rio de Janeiro, August 1982, p. 40.
31. See Lieber, 'Energy and Foreign Policy', p. 34.
32. This is why the buying-power wedge is a positive component of the import reduction premium, but a negative component of the oil stockpiling premium. See Plummer et al., pp. 22f. The annual cost of a stockpiling policy is the sum of physical storage costs plus the cost of tying up capital in idle inventories of oil (purchased at a higher than normal cost).
33. Hanns Maull, 'Time Wasted: The Politics of European Energy Transition', in Goodman et al., p. 275.
34. Krapels, p. 2.
35. See ibid., pp. 37ff; James Plummer and John P. Weyant, 'International Institutional Approaches to Energy Vulnerability Problems', in Plummer, pp. 260ff; and Louis Turner, *UK Interests and the International Energy Agency* (London, British Institutes' Joint Energy Policy Programme, Energy Paper No. 4, 1982), pp. 6ff.
36. The most important stipulations have been described in Krapels, pp. 37f. See also Turner, pp. 9ff.
37. Plummer and Weyant, p. 263, quoting from Chapter III of the agreement.
38. See Krapels, pp. 95ff.
39. Ibid., p. 4.
40. For a discussion of the various instruments which are available for these purposes—levying tariffs or higher taxes on petroleum products, providing subsidies to the domestic production of energy, and so forth—see, for example, Charles K. Ebinger, *The Critical Link: Energy and National Security in the 1980s* (Cambridge, Mass., 1982), pp. 39ff.
41. See Plummer, 'Energy Vulnerability Policy', pp. 22ff.
42. Ibid., p. 33.
43. Ibid., p. 34.
44. Lieber, 'Middle East Oil', p. 48.
45. See Lieber, *The Oil Decade*, pp. 6f, on which this section is largely based.

46. Plummer, 'Energy Vulnerability Policy', p. 4.
47. An especially interesting variant of this literature is Aaron Wildavsky and Ellen Tenenbaum, *The Politics of Mistrust* (Beverly Hills, 1981).
48. Krapels, p. 91.
49. Plummer, 'Conclusions', p. 339.
50. See Joseph S. Nye, David A. Deese and Alvin L. Alm, 'Conclusion: A US Strategy for Energy Security', in Deese and Nye, pp. 391ff; Ebinger, pp. 173ff; Daniel Yergin, 'Crisis and Adjustment: An Overview', in Daniel Yergin and Martin Hillenbrand (eds.), *Global Insecurity: A Strategy for Energy and Economic Renewal* (Boston, 1982), pp. 1ff; Robert J. Lieber, 'Cohesion and Disruption in the Western Alliance', in ibid., p. 320; and Krapels, pp. 115ff.
51. Deese and Miller, p. 201.
52. Plummer, 'Energy Vulnerability Policy', p. 1.
53. Lieber, 'Energy and Foreign Policy', p. 32.
54. Deese and Miller, p. 204.
55. Wilson, 'World Politics', p. 144.
56. Krapels, p. 113.
57. See Lieber, 'Cohesion and Disruption', pp. 327ff.
58. For two recent and contrasting interpretations of the role of the US in the creation of the IEA, see, respectively, Robert O. Keohane, *After Hegemony: Cooperation and Discord in the World Political Economy* (Princeton, 1984), pp. 217ff; and Robert J. Beck and Paul F. Jean, 'American Hegemonic Leadership and the IEA', paper presented to the American Political Science Association convention in Washington, DC, August 1986. Keohane views the creation of the IEA as the best current example of 'cooperation "after hegemony"', while the two latter authors question this judgement, and instead aver that it constituted a clear example of traditional American hegemonic influence (i.e. successful arm-twisting).
59. See Krapels, p. 114.
60. Ibid.
61. See for example, Lieber, 'Cohesion and Disruption', pp. 330ff.
62. Turner, pp. 44f. For recent US energy policy in general, see Hans Jacob Bull-Berg, *American International Oil Policy: Causal Factors and Impact* (London, 1987), especially pp. 170ff.
63. Yergin, p. 8.
64. See, for example, the figure in Bull-Berg, p. 173.
65. Ibid., pp. 8f.
66. See, for example, Knut Anton Mork, 'The Economic Cost of Oil Supply Disruptions', in Plummer, pp. 83ff; and Barry B. Hughes, 'The First Two Oil Shocks: Policy Response and Effectiveness', *Policy Studies Review*, vol. 5, no. 4 (May 1986), pp. 722ff.
67. Yergin, p. 9.
68. Quoted in ibid, p. 110.
69. Ibid., p. 7. This is Bull-Berg's recent characterization of the Reagan policy on energy, to distinguish it from the previous policy of 'circumscribed liberalism'. See his discussion of these two policies in Bull-Berg, pp. 170ff. He also has noted, however, that US policy has essentially been market oriented since the very beginning of the oil crises. Bull-Berg thus writes that 'From the start, the insurance role as opposed to a market interventionist part for the IEA was the

prime goal of the American IEA policy. The United States opposed conferring supranational powers on the IEA, it advocated an information-sharing scheme which sacrificed national security premises to the benefit of company integrity, and it was wary about regulating the oil price in an emergency situation. With Reagan in office the pro-company policy was enhanced as illustrated by the all-out opposition to price controls on crude allocated under the Emergency Sharing System of IEA' (p. 180).

70. Geoffrey Kemp, 'Military Force and Middle East Oil', in Deese and Nye, p. 368; see also pp. 365ff.
71. Ibid., p. 387.
72. Henry S. Rowen and John P. Weyant, 'Trade-Offs Between Military Policies and Vulnerability Policies', in Plummer, p. 313; see also 305ff.
73. Ibid., p. 322.
74. Bull-Berg, p. 182.
75. Hughes, p. 741.

3 Sweden and the Energy Crises

Historians of the next century may look back to the key period of 1970–80 and try to find out why Sweden and other major importing countries did not take stronger actions to counter the energy-vulnerability problem. If they read the pages of *Dagens Nyheter*, *Svenska Dagbladet* and *Veckans Affärer*, they will see an odd juxtaposition. On the front pages of these publications, they will see articles dealing with wars and instabilities in the Middle East, fuel costs driving up inflation rates, gasoline lines, coal strikes, nuclear-power protests, OPEC meetings and oil company profits. Buried on the inside of these publications, they will find articles dealing with the dominant energy policy issue of the era—whether this or that synthetic fuel technology would get government research funding, whether the breeder reactor programme would move forward or not, whether solar-related technologies would supply 5 per cent or 30 per cent of Swedish energy in the year 2000, or the long-term potential of gasohol from wheat grains. These energy historians may well puzzle over the mismatch between the short-term time-frame of the energy-vulnerability problem and the long-term time-frame of Swedish policy effort and expenditure.

The above paragraph constitutes a verbatim excerpt from the Stanford report on energy vulnerability referred to previously, except that the appropriate Swedish references have been substituted for the original American ones.[1] Such a substitution is illuminating for at least the following two reasons.

First of all, had I not immediately admitted to this subterfuge, probably very few people knowledgeable about Swedish energy policies would have noticed anything particularly untoward about the above claims.[2] In other words, Swedish and American responses to the energy crises of the 1970s seem to have had much in common, despite the fundamental differences which characterized their respective energy situations and policy commitments—or lack thereof—at the outset of the decade. Second, while the reference above to future historians is perhaps intended to suggest that only hindsight will reveal the implied mismatch between the vulnerability problem and policy responses to the energy crises, the original passage itself—and the massive study underlying it—effectively squashes any such assumptions. On the contrary, it is an obvious fact that in the United States the problems surrounding the short-term consequences of energy vulnerability have been thoroughly analysed and discussed at least by energy analysts and academic scholars. With respect to Sweden, however, recourse to future historians rather than to contemporary research and public discussion may very well turn out to be the only feasible option. Sweden has invested massively in energy research during the past decade—indeed, more than any other IEA country in per capita terms.[3] But I do not know of a single substantive study committed to analyzing the short-term effects on her national security and foreign policy of her enormous dependence on imported oil. Instead, we have been inundated with a plethora of research reports on everything from the quality of thermodynamic energy, the capital cost of aerogenerators to the availability of residual wood for biomass production.[4] A similar picture is provided by the state of public discussion on the energy problem. For most of the relevant period it has been almost totally preoccupied with nuclear energy rather than with broader national security issues.

These similarities and differences between the United States and Sweden provide us with a good starting-point for discussing some of the major problems which, in my view, have characterized the latter's responses to the energy crises of the period in question. The similar nature of the overriding issues involved affords us the opportunity of addressing energy vulnerability as a general problem, while the differences suggest that we look at Sweden as a case which needs to be analysed *sui generis*. Since we have already discussed some of the central aspects of the former, we can leave it aside for the time being and turn to an empirical consideration of the latter.

THE DEVELOPMENT OF ENERGY POLICIES, 1973–83

The fact that Swedish research and public debate on energy issues has been so weak with respect to the connection between her national security and energy vulnerability may, of course, reflect a condition which justifies such benign neglect. It may conceivably be the case that, objectively speaking, Swedish national security during the past decades has for all practical purposes remained an unproblematic issue, despite the convulsions on the international energy market; and hence, that Sweden has been in no real need of a major crisis-management apparatus in the event of new oil price shocks. This is obviously a central query of this study, especially in view of a widely held belief that Sweden's recent energy policies have been amongst the most responsive and successful in the industrial world. We will pursue this topic in the present chapter in terms of the development of Sweden's actual energy policies during the period under review, while the subsequent chapter will be reserved for an analysis of the logic of Swedish neutrality—the fulcrum around which Sweden's policy of national security revolves. Finally, in the concluding chapter we shall bring these two policy areas together in order to assess Swedish energy vulnerability as a crucial foreign policy and national security issue. This three-step approach is intended to facilitate an evaluation of the purported mismatch between Sweden's recent energy policies and the assumptions underlying her foreign-policy neutrality.

Historical Background

It is often assumed that prior to the first oil crisis the nature and implications of energy vulnerability were largely unknown as political issues, and hence that the immediate consequences of this crisis took the industrialized states by complete surprise. However, this is not entirely true, at least not for Sweden. On the contrary, this problem had been on the Swedish political agenda during at least two discernible periods prior to the 1970s: around the turn of the century, when the necessity of importing increasing amounts of coal to substitute for indigenous wood created both balance-of-trade and dependency problems; and during the 1950s, when oil first became a major source of energy. Indeed, it has even been maintained that not only was the issue of energy dependency not new when the first oil crisis hit Sweden, but that the way in which it was perceived and formulated at that time bore many of the characteristic hallmarks of discussions pursued earlier during the present century.[5] The gist of these can be found already in a motion submitted to the Swedish Riksdag (Parliament) in 1900, in which the economic and security aspects of Sweden's dependence on imported coal were described presciently as 'a Damoclean sword dangling over . . . [her] national independence and economic well-being'.[6] In a government study commission report on energy, published in 1956, this insight is specified further and in terms which have remained largely accurate since then:

> The crucial problem with the international oil market is its extraordinary sensitivity to political developments. The risk of disruptions due to war or threats of war is obvious, and it is increased by the fact that the Middle East—with its abundance of oil and fuel transportation networks—constitutes a focal point for international tension.[7]

As a result, attempts were made during both of these periods to generate and implement policies which would guarantee Sweden access to secure, domestically available and lasting sources of energy and energy-related technology. During the first decades of this century the availability of Swedish peat and hydroelectric power—the 'white coal', the 'patriotic fuel'—became the favoured long-term solution to this problem, while in the 1950s various combinations of indigenous energy sources, active conservation measures and domestically based nuclear energy were advocated in various study commission reports and policy statements.[8] The importance of research was also stressed in these earlier discussions of the energy problem—particularly, during the latter period, research on the use of atomic energy.

However, these perceptions of energy dependency would not continue to retain their urgency, especially not after the use of nuclear energy for peaceful purposes was given a decisive impetus by the 1955 conference on atomic energy in Geneva, when the nuclear powers decided to release large amounts of previously confidential information on nuclear technology. Since Sweden has the largest uranium reserves in Europe, the impact on her energy policy of this event was both swift and—it would turn out in due course—politically momentous.[9] Instead of the previous emphasis on the importance of pursuing active conservation policies, the potential future availability of this new source came to dominate an increasingly optimistic energy scene during the ensuing years. Or as an influential report on Swedish energy policies has put it:

By the time the Fuels Commission published its final report in 1956, all thought of pursuing an active conservation policy had been forgotten. The dependence on imports would not be solved by the exercise of restraint on the user side but by resort to nuclear power, that is, with measures on the supply side.[10]

And although in the beginning the intention was to create a nuclear energy system based entirely on domestic sources of uranium and knowledge—the so-called 'Swedish line'—the technical and economic problems associated with heavy-water reactors turned out to be so intractable that it is generally acknowledged that the programme itself turned out to be a complete flop.[11] Instead, the Swedish government was obliged to accept the adoption of light-water reactor technology, based on commercial American know-how and dependent on the import of enriched uranium.

Various explanations have been offered for this slide towards supply-side policies during the latter part of the 1950s. One major reason has already been mentioned: the emergence of domestic nuclear technology as a potential alternative to imported oil. It took a number of years until it became altogether clear that such a substitution would at best only be partially feasible (nuclear energy is primarily a substitute for hydroelectric power, not petroleum), and furthermore, that reactors would themselves be as dependent on the availability of imported sources of energy as systems based on fossil fuels. It is also obvious that once the giant oil fields of the Middle East began to produce its exceedingly cheap oil in volume, it was very difficult to resist becoming increasingly dependent on what was by far the most inexpensive and accessible form of energy around. Presumably it was also easier for the government to justify and administer an active policy for increasing supply rather than one aimed at decreasing demand, especially during a period characterized by the public's growing confidence in the ability of the market to provide not only a constantly rising standard of living but also the necessary wherewithal for regulating the energy sector. Thus, for example, although the government had at first intended to exercise a decisive influence on the development of the nuclear energy sector (partly in order to keep open the option of producing nuclear weapons), the problems associated with heavy-water technology and the concomitant decrease in the price of petroleum products led to a step-wise weakening of its political will in this regard.[12] Finally, as Måns Lönnroth and his colleagues have noted in this matter:

> decisions which sometimes have crucially affected the course of events have been made outside (or rather below) the political level in society. Such decisions may have to do with rate structures, satisfaction of engineering criteria, forms of collaboration and joint ventures between, say, the producers of electrical power and its distributors. Trade associations and professional groups apparently play big roles. To all intents and purposes, therefore, the practical scope for assigning political accountability for the structure of the energy supply system is very limited indeed.[13]

Whatever the reasons, however, the choices made between the 1950s and the early 1970s were successful in one crucial regard: they provided Swedish society at large with very cheap and readily available energy during a period of unprecedented economic prosperity. Indeed, the latter would probably not have been feasible without the former.

A New Energy Strategy

Two empirical factors contributed to change Swedish energy policies fundamentally during the 1970s: the oil crises (especially the first of these) and the growing public opposition to the nuclear power orientation of the Swedish energy system. Although these developments coincide in time and to a considerable extent overlap as substantive concerns as well, we shall here review them as separate political issues.

The immediate Swedish reaction to the oil embargo in October, 1973, was to shift 15 per cent of the total supply of oil from household to industrial use, and to set in motion a crisis energy conservation programme which included fuel rationing for private automobiles for three weeks in early 1974, information campaigns directed at both industrial and household users, and governmental subsidies to companies for training rather than laying off employees. As a result, the industrial sector succeeded in reducing its demand for oil by 5 to 10 per cent, while households reached figures of about 15 per cent. Fortunately, the crisis was too short to cause any immediate substantial effects on employment and production, although some large companies—such as Volvo—were forced to shorten the work week for several weeks.[14]

Instead, it is generally acknowledged that the immediate impact of the embargo was on public attitudes rather than on the economy: for the first time it became palpably obvious to everyone how extremely vulnerable Sweden in fact is to peacetime events beyond her borders and her control. Leon Lindberg has expressed this change in attitude as follows:

The behaviour of the Swedish energy policy system in response to the events of 1973–74 is distinctive in the following ways: Alternative policy criteria penetrated policy debate at the highest political levels; the mass public has been mobilized and made aware of the larger stakes of energy policy; and decision making at the bureaucratic-technocratic levels has been made permeable to new actors and new perceptions of the problem.[15]

One result of this upsurge in public interest was a government initiative in January 1974, aimed at creating study groups on the energy issue throughout the country. These groups were organized under the aegis of voluntary adult education associations, which received special state grants for this purpose. Altogether about 8,000 such groups were set up, involving some 80,000 participants. This activity led to an intense public debate on two basic energy issues, the one related to supply, and other to demand: the type of energy which Sweden should use, and how much energy she really needed. However, as Lönnroth has noted, these discussions 'rapidly evolved into a debate based on fundamental political views of society', for example the relationship between energy supply and such social goals as full employment, higher levels of welfare, environmental improvement and so forth.[16] These more wide-ranging queries were not raised by the oil-dependency issue alone but also by questions relating to nuclear energy which—it must be emphasized—had been initiated already prior to the oil embargo.

One immediate consequence of this intense involvement with energy issues was a clear-cut shift away from an emphasis on the production of electricity to a policy stressing conservation and the development of renewable energy sources. As several commentators have emphasized, this response was not typical for what Lindberg, in an influential book of the 1970s, has diagnosed as the 'energy syndrome'.[17] This term refers to a series of interrelated symptoms describing a system malfunction in the attempts of industrial societies to respond to rising energy prices and the long-term implications of the eventual depletion of certain energy sources. These include the following:

> production and consumption practices that require steady increases in energy supply; public policies that are dominated by the perspectives of energy producers and that are almost exclusively oriented to meeting exogenously determined demands for energy; and an interacting set of political, institutional, and structural obstacles that constrain the search for alternative policies.[18]

In his recent book on domestic Swedish energy policies, Robert C. Sahr has argued that Sweden is a partial exception to this syndrome, particularly with reference to her early recognition of the importance of demand policies, but also in regard to early Swedish interest and investment in alternative energy sources.[19] The first of these shifts is most clearly evident in the overall energy policy adopted by the Swedish government in May 1975, and subsequently amended several times, while the latter is primarily indicated by the scope of Swedish research and development efforts within the energy field. On the following pages we will focus, in some detail, on the nature and implications of these two shifts during the period considered in this study. We will also, given the main thrust of our analysis, consider how Sweden's oil stockpile policies have developed during the years in question.

A Comprehensive Energy Policy

(i) The decision on a new and comprehensive energy policy taken by the Swedish Riksdag in the spring of 1975 has been regarded as the definite end of one era and the beginning of another. During the period thus brought to a close, energy was basically a peacemeal issue handled—when the need arose—on the administrative level, and only rarely discussed by either the government or the Riksdag. This conformed to a situation in which the availability of electricity was the state's main concern in this area, while the supply of oil was left wholly to market forces and the self-adjusting structural and price mechanisms underlying them. The direction which the new era was to take can be gauged from the four 'cornerstones' of this policy, presented to the Riksdag by the Social Democratic government in May 1975:

To decrease the growth rate of energy demand. The intent was to restrain consumption to an average of 2 per cent of energy demand growth between the years 1973 and 1985, from the previous average annual rate of 4.5 per cent. A series of measures were proposed to achieve this goal, including increased taxation of energy, grants for retrofitting in space heating and for the increased use of waste heat from industrial plants, and so forth, as well as investments in an extensive research and development programme, divided about equally between the supply and demand aspects of energy policy.

To pursue an active oil policy. This included a reduction in overall demand of oil; the increased import of other fossil fuels, especially coal and natural gas; increasing the oil stockpile; prospecting for oil in the Baltic Sea and creating a state petroleum company for trading in crude and petroleum products; developing bilateral agreements with the oil producing states; and cooperating with other oil importing countries.

To increase the supply of electricity. On the assumption that demand for electricity would increase relative to demand for oil, the intent was to invest in two additional nuclear plants (to be added to the eleven previously approved), as well as in some further hydroelectric power.

To partake actively in international cooperation. This included membership in the IEA (with a special agreement in consideration of Sweden's neutral status), as well as active participation in alleviating imbalances between the rich and poor oil importing countries of the world.[20]

Let us look at selected aspects of this policy in terms of the main concerns of this study. In the present chapter we will concentrate on its domestic characteristics, leaving the international aspects for Chapter 5.

It is in particular the first of these 'cornerstones'—the initiation of a concerted effort to restrain consumption—which deviates from the policies pursued by most other comparable nations at this time. Sahr has thus noted that whereas the latter responded to the energy crisis with at least a rhetorical emphasis on the need to 'save' energy, Sweden 'appears to be the earliest nation to make energy conservation a central part of its energy policy, at least the equal of policies oriented toward strengthening energy supply'.[21] More specifically, the policy specified that affecting energy demand would henceforth be a goal equal in rank to that of assuring supply. In the language of energy economics, this meant the abandonment of the traditional practice of allowing energy demand to be an exogenously determined variable in energy policy.

One surmisable reason why a serious commitment to conservation was internationally not more common at this time is that it is a policy which, for basic structural reasons, is far from easy to implement. Shifting from a supply orientation to a policy of deliberately managing both supply and demand constitutes—both conceptually and organizationally—a fundamental change in the energy system of a society. Thus, for example, whereas a supply-oriented policy is directed toward a determinable number of actors (in the case of Sweden, between ten and twenty private and public enterprises), this does not hold for demand-side policies, which of necessity involve a very large number of decision-makers—from large cities and companies to single individuals—dispersed across all levels of society. Furthermore, while the primary aim of the former is well-known and clear-cut, namely to produce and sell energy in one form or another, this is obviously not the case with respect to the latter, for whom energy is hardly ever a goal in itself but simply a means for achieving something else, be it the production of trucks, the transportation of goods or making ice-cubes. Hence, whereas the former are in principle relatively easy to influence or regulate, the opposite holds for the latter, for whom energy in each instance is merely part of a larger process of purposive behaviour. Consequently, as Lönnroth has written, 'there are reasons to believe that energy conservation policies interact in a very complex way with other policy areas, and sometimes in a rather unpredictable manner'.[22] This obviously complicates the planning of measures afflicted with a minimum of unintended and undesirable consequences.

To this conceptual complexity must also be added the organizational problem of channelling decisions within a system traditionally lacking in the necessary administrative infrastructure and professional incentives for balancing supply and demand in this deliberate fashion. This problem has at least two aspects which are relevant here. First of all, the energy policy substructures of most Western industrialized states have traditionally been interdependent systems in the technical sense, while at the same time being exceedingly fragmented with respect to their organizational and institutional characteristics. Thus, for example, while technologically the present Swedish electricity system forms an interconnected network of hydropower stations, nuclear reactors, power transformers, national distribution grids, household meters and so forth, the corresponding organizational picture bears little resemblance to this type of interlocked, concatenated system. Rather, it is handled by a host of different and independent organizational units, both public and private; and these are, in turn, regulated or influenced by a large number of legal, financial, labour market and other norms, regulations and expectations, many of which have evolved or been created in complete disregard—or even ignorance—of one another.[23] This has resulted in an administrative complexity which places very obvious structural constraints on the successful implementation of conservation policies.

Second, policies of demand restraint are not easily imposed on an energy industry and an energy bureaucracy whose members have traditionally thought of policy exclusively in terms of increasing supply rather than reducing growth rates. In other words, it requires a considerable change in familiar ways of thinking on the part of two of the most central policy actors on the Swedish energy arena. More specifically, as Sahr has suggested, a serious commitment to a policy of demand reduction 'could change the relationship between the energy bureaucracy and the energy supply sector from one of subsidization and encouragement to one of regulation'.[24] As we know from Theodore S. Lowi's analysis of the implications of moving from distributive to regulatory and redistributive policies, this is one of the most difficult types of changes to implement without strong resistance from the actors affected by it.[25]

Despite these various problems, the demand restraint policies proposed by the government and approved by the Riksdag in 1975 were intended to be more than symbolic; and they were on the whole accepted by the actors most directly involved in this policy area, that is the energy producers (both public and private) and the Ministry of Industry, the main public bureaucratic actor in Swedish energy policy. Various explanations have been offered for this relative success of the government's proposals. The Ministry of Industry, for example, did so because it was 'able to affect the direction of the policy in ways that helped preserve established practices and arrangements while at the same time allowing for change', according to one knowledgeable scholar.[26] In other words, it was able to accept the necessity of a policy of conservation because doing so did not mean having to forfeit much of its traditional control over the energy policy area. As to the energy producers, the following factors have been adduced to explain why resistance to the proposed changes was less than might have been expected.

First, the various groups within the energy sector saw some advantages in reducing the rate of energy growth. Second, the government energy officials were able to

> reassure those within the private sector that concern for energy conservation would not be allowed to undercut ongoing economic goals and commitments. Third, there was some vagueness built into the policy, or rather left undefined, which allowed for future modifications of policy mechanisms and goals for energy conservation and which prevented closing of options.[27]

Also, since the figure of 2 per cent per year was specified as an average for the whole period 1975–85, it left open—except in one respect, to which I shall return below—the question as to how, in immediate policy terms, this goal was to be achieved. Consumption could thus be allowed to increase substantially more than 2 per cent during the initial years of the decade in question without in any sense contravening—either in letter, or so it would seem, in spirit—the stipulations of the 1975 decision on energy policy.[28]

This was, in fact, a conscious policy on the part of the government, as Evert Vedung has noted. 'The Social Democrats', he thus writes, 'explicitly expected that the rate of increase during the first years would lie significantly over the average—perhaps at 3.5 to 4 per cent—before settling around 1 per cent toward the end of the period'.[29] In other words, the fact that the 2 per cent goal was formulated as an average for the entire period meant that no decrease in energy consumption was called for until 1978. This allowed for the possibility (e.g. if a change of public opinion occurred) of introducing an upward adjustment of the consumption rate at a later date.

The exception noted above pertains to the relationship, with respect to the annual rates of demand restraint to be achieved by, respectively, the industrial, the transport and the residential/commercial sectors. The intention was that the first should be least affected by the new policy (3 per cent increase per annum), the second somewhat more (2 per cent), while the largest savings in energy use—no annual increases at all—were to be achieved in homes and offices (mainly through efficient space heating). This relatively liberal attitude towards the needs of the industrial sector obviously also helped to assuage some of the misgivings prevalent within it.

While the government's conservation policies were given an explicit numerical target in the 1975 decision, this was not the case with respect to its policy on petroleum substitution. In this matter Vedung, for example, speaks of the presence of only 'scattered indications' in the government bill, and comes to the conclusion that the consumption of petroleum was not intended to increase by more than 1 per cent per annum during the period 1975–85.[30] In other words, oil consumption is assumed to increase during the period in question, but only marginally. More specifically, this conforms to a policy, in a situation of annual increases in energy consumption, of allowing oil consumption to increase in absolute figures, while decreasing it in relative terms, that is in terms of the total consumption of energy. This is also evident from Table 3.1, which shows that while the use of oil is allowed to increase in absolute numbers between 1973 and 1985 (from 313.3 TWh to 324 TWh),

Table 3.1 Types of energy and consumption

Type of energy	Actual consumption 1973		Projected consumption 1985	
	TWh	%	TWh	%
Oil products	313.3	73.1	324	60.0
Coal, coke	19.5	4.5	50	9.0
Lyes, firewood, waste	33.7	7.9	37	7.0
Hydroelectric power	59.9	14.0	66	12.0
Nuclear power	2.0	0.5	63	12.0
Total energy supply	428.4	100.0	540.0	100.0
Average annual change			+2%	

Source: *Prop. 1975:30*, Appendix 1, p. 358 (1973) and p. 363 (1985).

the 1975 energy decision projects a decrease, during the same period, of over 13 percentage points when account is taken of oil's share of the total energy consumption.[31]

The aim of decreasing the relative consumption of oil could in principle be put into effect in two ways: oil could be replaced with other *fuels* such as coal, coke, wood chips or natural gas; or it could be replaced by *electricity*, especially in the case of space heating. If we look at Table 3.1 again, it is evident that Parliament had both of these options in mind in its 1975 decision, but that electricity was given the major role in its oil substitution policy. Vedung here speaks of 'a clear intention on the part of Parliament to encourage continued increased use of electricity'; and in the bill itself a figure of 6 per cent per annum is suggested.[32] Indeed, the intention was to double (in absolute terms) the consumption of electricity within little more than a decade. Furthermore, while electricity can be produced in oil-fired power stations, it was clear, given the general thrust of the new energy strategy in the wake of the first oil crisis, that this was not what the Riksdag had in mind. Rather, the greater part of this increased supply of electricity was to be provided by uranium-based nuclear power plants (Table 3.2).[33]

(ii) The decision made by Parliament in 1975 was amended when, in May 1978, it adopted the new Fälldin government's first major energy bill, specified in terms of a conservation plan for existing buildings. This bill stipulated that within a decade (i.e. by 1988) the net consumption of energy within this sector should be reduced by approximately 35 TWh per annum.[34] This target was specified for *all* existing buildings, including those belonging to the industrial and commercial sector. Thus, although it is difficult to compare the 1975 and 1978 decisions, since they are expressed in terms of different operational criteria (percentages versus TWh) and different—though overlapping—consumption categories (socioeconomic sectors versus all existing buildings), it is nevertheless fair to claim that the goals

Table 3.2 The production of electricity

Electricity source	Electricity use in TWh in 1973	Projected use in TWh in 1985
Hydroelectric power	60	66
Nuclear power	2	63
Industrial backpressure	3	8
Combined heat and power	4	15
Other thermal power	9	7
Total energy supply	78	159
Annual average change		+6%

Source: *Motion 1965:2030*, p. 13 (1973 figures) and *Prop. 1975:30*, Appendix 1, p. 362 (1985 figures).

contained in the Fälldin bill were more ambitious than those for corresponding areas in the previous Social Democratic policy.

(iii) When adopting the 1975 energy programme, the Swedish Riksdag also decided to review these comprehensive energy policies by 1978. However, it could not hold to this timetable, mainly as a result of the increasingly divisive character of the nuclear energy issue (see below). In fact, since the first Fälldin government—consisting of the three non-socialist parties which, in 1976, had won its first victory over the Social Democrats for many decades, partly as a result of the nuclear issue—was dissolved in October 1978 due to this very issue, Parliament was not presented with a review opportunity until 1 March 1979, when the new minority Liberal government headed by Ola Ullsten submitted a plan for a revised comprehensive energy programme. However, as a consequence of the Three Mile Island accident on 28 March, and the subsequent decision (on 4 April) to hold a referendum on the future of nuclear power as a source of energy in Sweden, only selected parts of this programme were in fact considered and adopted.

On the whole, this new programme entailed a strategy for reducing energy demand along the same lines as the previous policy, except that the scope of savings envisaged is even more ambitious than in previous decisions. For the industrial sector, an increase in consumption for the period 1978–85 of between 2.7 and 3.4 per cent per annum was indicated, and between 1.8 and 2.4 per cent per annum for the period 1978–90. No numerical targets are given for the transport sector, but a slight annual increase in consumption is envisaged. The most substantial savings are still to be made within the residential commercial sector: a *decrease* of between 0.8 and 1.8 per cent per annum for the first period, and approximately −0.7 and −1.6 per cent per annum for the period 1978–90. These targets are well below those suggested in the 1975 energy plan, as is the overall figure for the year 1985. Thus instead of the 500 TWh rate for 1985 contained in the 1975 programme, the

1979 bill speaks of a rate of consumption within the 425–50 TWh bracket, that is an additional reduction of between 50 and 75 TWh by 1985.[35] This ambition is also reflected in the bill's projection of an increase in the demand for energy between 0.4 and 1.0 per cent per annum for the period 1978–90, as compared to the 2 per cent figure posited in the 1975 energy programme.[36] Since the rate of increase for the member states of the IEA for the same period had been set at approximately 2.8 per cent per annum, this plan was regarded by the government as a considerable achievement in projected demand restraint. However, due to the reasons mentioned above, this target for a comprehensive energy goal was one of the sections of the bill which was not considered by Parliament.[37]

For the first time in Swedish energy policy, this bill contained the assumption that the import of oil would decrease not only relative to other energy sources but also in absolute terms. However, oil continued to constitute close to 70 per cent of the available energy supply in 1978, a figure which obviously remained highly unsatisfactory, since oil substitution was still regarded as the major thrust of Swedish energy policies for the whole period extending to 1990.[38]

(iv) The comprehensive energy goals set by the Riksdag in 1975 were not formally changed until May 1981, when the second Fälldin government (consisting, once again, of the three non-socialist parties, which had won the election in 1979) submitted a new energy policy programme to Parliament. In line with the ever-increasing downward trend characterizing the consumption of energy during this period (mainly due to the second oil crisis and the stagflation characterizing the international economic system as a whole), the projected rates for the 1980s were lowered even further. Indeed, for the period extending to 1985, Parliament now decided on a rate extending from −0.7 to 0.6 per cent per annum, that is a figure averaging around zero. For the period 1979–90, the comparable figures are −0.4 and 0.3. Thus, while in the 1975 bill a situation of zero growth in energy consumption was not envisaged until after 1990, we here have a projection of this state of affairs which had been brought forward by virtually a whole decade.

The differential rates for increases within, respectively, the industrial, transport and residential/commercial sectors are retained in the 1981 bill, although an even sharper distinction is made between the needs of industry—which is to receive all the energy that it needs—and those of the residential/commercial sector, where the biggest savings still are to be made: within a range of −3.2 and −2.1 per cent per annum for the 1978–85 period, and −2.4 and −1.4 for the whole period 1978–90.[39]

Consonant with the Social Democratic view contained in the 1975 decision, electricity continues to be regarded as the main substitute for oil. For although the projected 1985 consumption of electricity was considerably lower in the 1981 bill as compared to the 1975 decision (109 TWh, compared to 160 TWh, the latter including losses through energy processing), elec-

Table 3.3 Energy consumption 1973–83

Sector	1973		1979		1983	
	TWh	%	TWh	%	TWh	%
Industry	162	41	155	39	131	37
Transport	65	17	70	17	68	20
Commercial/residential	163	42	177	44	150	43
Total domestic consumption	390	100	402	100	349	100

Source: *Prop. 1984/85:120*, p. 43.

tricity produced by nuclear power plants was to remain—at least during the 1980s—the major instrument for combating the very high oil dependency still characterizing Sweden at this time. In addition, however, the 1981 decision also advances the role of other fuels, especially coal, as a substitute for oil.[40]

The 1981 energy bill is the last major policy decision made by Parliament for the period covered by this study. In 1982 the Social Democrats succeeded in returning to power once again and presented their first energy programme in early 1985, when much of the crisis atmosphere characterizing the energy scene of the 1970s and early 1980s had dissipated.[41] Part of the reason for this can perhaps be gauged from the following facts, which can also serve to round off this discussion.

First of all, as Table 3.3 shows, the total consumption of energy per annum had declined from 390 TWh to 348 TWh, that is by approximately 1 per cent per year.

For the industrial sector, the use of energy had declined despite an output which had remained essentially stable during the whole period. More importantly, the role of oil within this sector had decreased considerably, namely by about 6.3 per cent per annum. Although the transport sector had not succeeded as well, the opposite was the case with respect to the commercial/residential sector which, despite the inevitable expansion of its physical size during a ten-year period, had lowered its consumption of energy in absolute terms from 163 TWh to 150 TWh. Furthermore, while oil had accounted for about 90 per cent of this sector's energy consumption in 1970, this figure had dropped to 40 per cent by 1983.[42] Even more encouraging are the figures for the relative distribution of the types of energy supplied, as indicated by Table 3.4.

Oil, which had accounted for over 70 per cent of the energy supply in 1973, and remained slightly below this figure until 1979, had now dipped to 53 per cent. Even more significant was the fact that this downward tendency was showing no signs of levelling off. A major reason for this decline—considerably greater than envisaged in the 1975 decision—was the successful expansion of electricity during this period: having represented about 15 per cent in 1973, it now accounted for a full quarter of the nation's energy supply. This

Table 3.4 Total supply of energy 1973 and 1983

Type of energy	1973		1983	
	TWh	%	TWh	%
Oil products	338	73	214	53
Domestic fuels	45	10	54	13
Coal	20	4	23	6
Hydroelectric/nuclear	62	13	110	27
Waste	—	—	2	1
Total supply	465	100	403	100

Source: *Prop. 1985/85:120*, p. 44.

development away from oil was due mainly to the expansion of the nuclear energy sector during the period in question, but also to the fact that while oil had accounted for approximately 40 per cent of the energy input in the production of electricity in 1973, the comparable figure in 1983 was a mere 5 per cent.

Energy Research and Development Policies

The second fundamental shift in Swedish energy policies pertains to the recognition, in the immediate wake of the oil embargo, of the need to invest substantially in a concerted energy research and development programme, especially with the view of discovering alternative energy sources. Though somewhat sudden, this was nevertheless a wholly logical insight on the part of the Social Democratic government, given the accumulation—during a few crucial years around 1970—of the multiple and reinforcing problems associated not only with the increasingly vulnerable nature of energy supplies dependent on Middle East sources but also with the *cul-de-sac* of the heavy water nuclear programme and the mounting political, environmental and economic complications attending the continued expansion of nuclear energy produced in other forms.

Late in the autumn of 1973 the government thus appointed a high-priority body—the Energy Programme Commission—entrusted with the task of making an inventory of ongoing research in the energy area, of specifying the major needs attending it and, most importantly, of formulating a comprehensive research programme specified in terms of priority-research problem areas. Within less than a year this well-endowed commission presented its main findings, and following the traditional *remiss* procedure, that is of circulating the report among relevant public and private organizations and institutions for their comments, these findings were, with minor changes, incorporated into the comprehensive energy policy bill passed by the Riksdag in the spring of 1975.[43] The programme was in the first instance to run for three years; and although the commission's recommendation of a 420 million SEK budget for this period was reduced to 366 million SEK, it still compared very favourably with the sum of 45 million SEK which had previously been allocated for this purpose for the fiscal year 1975/6.[44]

As we have already noted in passing, this was not the first public Swedish decision to invest in energy research, nor was it the first time that perceptions of vulnerability had led to political efforts to develop domestic energy substitutes for foreign imports. However, previous efforts had on the whole been modest, except in the case of the development of the 'Swedish line' of nuclear technology based wholly on the use of domestic uranium. This time around, however, it was no longer a question of yet another small-scale research investment aimed at alleviating a short-term supply problem. On the contrary, the research effort activated on 1 July 1975, turned out to be the largest civilian research and development ever to be launched in Sweden.[45]

A major reason why this programme could be established with so little political fuss or bureaucratic obstruction was the fact that it was not structured in terms of a single, hierarchical framework for initiating or implementing policy. Creating such a new organizational structure would probably have caused considerable conflict and delay, and was hence—given the perceived urgency of the energy issue—not seriously considered. Rather, instead of taking its cue from the sectorial research system which had become increasingly paradigmatic for Swedish science policy during this period (i.e. creating permanent research institutes and units wholly within the ambit of single ministries, whose policy needs they were in the first hand to serve), a diversified administrative structure was established. Hence, agencies within four different ministries were designated to act as planning and funding units, with the Ministry of Industry acting as co-ordinating centre for the multi-annual budgetary process. Furthermore, whereas some of these units, such as the National Board for Technical Development (STU) predated the energy programme and were to have energy issues as just one part of their regular activities, others—especially the Energy Research and Development Commission (DFE)—were specifically created to promote and monitor the interests of the new programme. It is in this sense that the energy research programme constitutes 'the main example in Swedish science policy of an effort to link a comprehensive mission-oriented research program to an organizational structure that cuts across sectorial boundaries', as Björn Wittrock and his associates have noted.[46]

Another reason why this programme could be launched with such dispatch was the existence, already prior to the commission's work, of an extensive if informal network which was not only available for eliciting ideas but many of whose members could subsequently also be utilized for carrying out large parts of the programme itself. Probably, this is also one of the reasons why Sweden was quicker—and perhaps also more successful—than most other countries in escaping at least some of the more insidious consequences of the 'energy syndrome' discussed above. This network consisted of industrial, governmental, academic and other research-oriented groups and individuals, with the Academy of Engineering Sciences (IVA) playing a crucial role as clearing house (as it had done for a number of years). These interested parties were in fact involved not only as members of the expert groups established by the commission but also in the actual drafting of the report itself. And since they—for example the Royal Institute of Technology in Stockholm and the large nuclear research institute in Studsvik—were to become the major beneficiaries of the programme thus created, we here have an example of the existence of 'a kind of implementation structure before implementation'.[47] It is also for this reason that Wittrock and his colleagues have viewed the Swedish energy research programme as a prime example illustrating the central role, in the process of actual implementation, of informal networks existing prior to the policies to be implemented.[48]

However, despite the crucial function of these informal networks, as well as the relatively decentralized character of the organizational structure of the programme, the recommendations of the commission were above all distinguished by its efforts to view energy issues in terms of an overarching, systemic and long-range perspective. In this sense it came to typify, at least on paper, the extremely rationalistic tendency which is associated with Swedish policy-making during the 1970s—'the era of large programmes', as Wittrock and Lindström came to call it in the title of their book.[49] This type of policy blueprint is premised on the assumption that implementation is a delegated task involving the purposive pursuit of previously established and unified programme objectives by means of a concerted, hierachical effort on the part of a well-structured, self-conscious and rational bureaucratic organization created or appropriated for the purpose.

Translated into the specific terms of the programme itself, this meant that a *Main Programme for Energy R & D* was established, consisting of six programmes.[50] With one exception, these are divided into a number of subprogrammes (see Fig. 3.1 for programme and subprogramme titles, and of the names of the agencies responsible for them).

The responsible agencies had to work out programme plans for all important areas of energy R & D as follows:

> These are formulated so as to focus the projects during the programme period on the objectives officially decided for the energy R & D programme, as well as to provide a basis for official decisions as to how energy R & D is to continue after the programme period. The programme plans are continuously revised by the responsible agencies.[51]

As already mentioned above, a special agency—the Energy Research and Development Commission (DFE)—was created, with the responsibility not only of co-ordinating the programme as a whole but also for providing information which could serve as a basis for the continuous overall planning of energy R & D. In addition, it was given the function of evaluating the results produced by both Swedish and international research efforts.

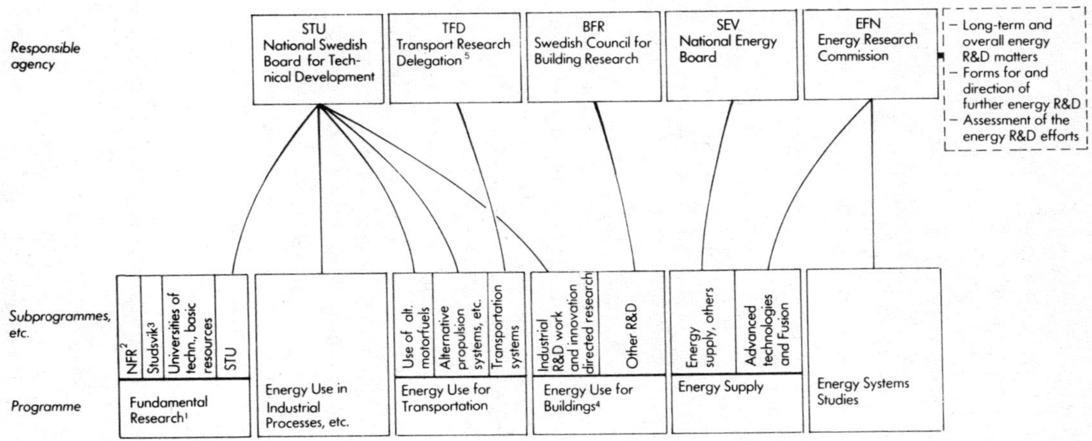

Notes:
(1) No agency has the overall responsibility for the whole programme
(2) The Swedish Natural Science Research Council
(3) Studsvik Energiteknik AB
(4) BRF is the responsible agency for the programme
(5) The Transport Research Committee has in its final report proposed that TFD should be replaced by a new agency, the National Board for Transport Research (from July 1, 1982)

Source: DFE Report No. 42 (1982), p. 26.

Figure 3.1 Energy Research Commission

Things did of course not always work out quite the way anticipated by the more optimistic proponents of this rationalistic blueprint. Especially the colleges of technology, which from the very beginning received substantial financial support from the programme, were often happy to do so without committing themselves deeply to programme objectives defined elsewhere. When large amounts of money trickle down to a number of highly individualistic R & D environments (it has been noted in this connection), 'the idea of some kind of unifying program rationale tends to get tenuous at best'.[52] On the other hand, the fact remains that the programme launched in 1975 was apparently flexible enough, despite its rigid systems management hauteur, to generate substantial amounts of the kind of research that it was intended to support. In a review of its development this is explained as follows:

> An administrative conception based on ideals of coordination, control and hierarchical management has proved to be quite compatible with an implementation process that is largely determined by the requirements of units at the bottom level and their normal operations. By allocating money with considerable discretion it has been possible for program agencies to gradually track down those R & D performing units that they deem worthy of further and enlarged support.[53]

Indeed, Wittrock and his colleagues have noted that in some important respects the energy programme has functioned better than similar projects initiated and run entirely by sectorial authorities.[54]

Officially, the specific objectives of this national research programme have varied somewhat since it was initiated, as can be gauged from the following three periodic specifications. In the commission report establishing the programme, its goals were described succinctly in terms of making the supply of energy in Sweden more efficient by acquiring knowledge aimed at (1) securing and increasing the availability of supply, (2) decreasing the need for energy and (3) decreasing the per-unit cost of energy consumption.[55] For the period 1978–81 the principal aim of the programme was described in terms of providing better means for coping with the following tasks:

- Increasing energy conservation.
- Creating new means of energy supply, especially by using domestic and renewable energy sources.
- Diversifying energy supply over different sources, primarily by the replacement of oil.
- Creating larger flexibility in the energy system.
- Producing technologies and methods that have a lower impact on the environment.[56]

Finally, in the report on the third programme (1981–4), the following objectives are specified:

The overall target for actions in the energy research programme involves concentrating the main thrust of effort on lines of development capable of leading to commercial exploitation and of providing valuable contributions to the energy balance in the short and medium term time perspective. To this end, the major share of the resources will be directed towards research and development work on conservation and on sources of energy capable of replacing oil and oil products. Concurrently, the technological development will be monitored in fields of potential long-term significance to our energy situation.[57]

Two main considerations have thus characterized the objectives of this extensive research effort: promoting the expansion of knowledge with respect to the energy field as a whole, including knowledge leading to the introduction of new technologies and new energy systems; and contributing support in the actual introduction of the new technologies that are required to enable Sweden to achieve the main thrust of her energy policies in general.[58]

An estimate by the IEA indicates that during the first period of the research programme, expenditures for energy R & D in Sweden increased by more than 250 per cent in real terms—the highest increase of all IEA countries. It is also interesting to note that during this period Swedish efforts were exceptional in another sense as well. Thus, whereas on the average only 10 per cent of energy research monies in the IEA countries were spent on energy conservation and renewable energy sources, in Sweden 30 per cent and 25 per cent, respectively, were allocated (during 1978–9) for these purposes. On the other hand, the proportions devoted to coal and nuclear technology were markedly lower in Sweden than was the case—on an average basis once again—for Western Europe and North America as a whole.[59]

As already indicated above, the energy research programme continued to run for new three-year periods following the end of programme discussed above. The contents of the second programme, covering 1978/9–80/1, was based on a study done by the Energy Research and Development Commission, and despite new directives from the new three-party government under Fälldin, this led to a decision by Parliament to continue the programme with little change, except that a greater emphasis was to be placed on research on renewable energy sources.[60] The new government was, however, eager to increase the level and scope of the programme (Fälldin himself had been very involved in the energy issue during his election campaign in 1975), with the result that close to 1,000 million SEK were allocated to its second three-year period.

For the third three-year period (1981–4), a budget of 1400 million SEK was approved by Parliament, which, once again, represents a significant increase over the sums allocated for the previous period (Table 3.5).

Starting with this period, the programme was to be financed by increasing the strategic reserve levy charged on certain oil products. In view of the need for more powerful co-ordination in the planning and implementation of

Table 3.5 Energy R & D in Sweden, three-year programmes, 1975–84 (millions)

Energy R & D category	1975–8 Skr	$	1978–81 Skr	$	1981–4 Skr	$
Energy in industrial processes	43.5	9.7	99.0	23.4	171.0	23.4
Energy in transportation	26.0	5.8	35.3	8.3	70.0	9.6
Energy in buildings	68.0	15.2	154.5	36.5	256.0	35.0
Energy supply	215.8	48.3	458.75	108.5	700.0	95.6
Energy systems studies	6.2	1.4	15.0	3.5	25.0	3.4
Fundamental research	16.8	3.8	30.5	7.2	50.0	6.8
For later allocation					113.0	15.4
Planning and coordination	6.0	1.3	12.0	2.8	15.0	2.0
Total	382.3	85.5	805.05	190.2	1,400.0	191.2
Inflation adjusted (1980 = 100)	ca. 510		805.05		ca. 1,140	

Source: Sahr, p. 130. See also *SOU 1986:31 EFU 87*, p. 19.

energy policies (especially following the referendum on nuclear energy, to be discussed below), a new and more centralized organization was also created. This new organization consists, in the first place, of a new government agency—the National Energy Board (SEV)—primarily charged with all matters pertaining to the supply and overall planning of energy. The Board has also assumed responsibility for short and medium term research, while a special authority—called the Energy Research Commission (EFN) and superseding the Energy Research and Development Commission—was established for pursuing long-term and overall energy research questions. This body has also been encharged with monitoring and evaluating the long-term, comprehensive aspects of energy R & D issues. A special agency was also created for administering the new funding system referred to above.[61]

In conclusion, it should perhaps also be noted that although Swedish R & D efforts within the energy sector during this period have been extraordinary with respect to the sheer size of the financial investments which successive governments have committed to it, these policies have at the same time followed a cyclical pattern which is by no means unique. Since Wittrock has recently given us an excellent description of this development, I shall here quote it *in extenso*:

During periods of great uncertainty and crisis-like events, the government has launched ambitious programs which have entailed fairly far-reaching control and intervention in the process of technology development and innovation. The main objectives emphasized have concerned national self-reliance and trade policy. The programs have typically involved substantial efforts to stimulate technologies which might tap domestic energy sources. These research and development programs have often had a long-range orientation and have rested upon assumptions of the future profitability in market terms of technologies under development. The state has also taken on the main responsibility in financial terms. However, the periods of crisis have tended to be of fairly short duration. When the immediate sense of uncertainty has waned, the political feasibility of far-reaching state intervention and control has also subsided.[62]

At such a point, some type of policy drift tends to set in, changing the basic thrust of a policy without necessarily curtailing its scope. Essentially, this seems to be the situation at present with respect to energy research policy in Sweden.

Oil Stockpiling Policies

The purpose of emergency oil reserves is to ensure that extra oil is available in the event of a cut-off in supplies. Such stockpiles differ fundamentally from commercial inventories, which exist as day-to-day buffers in the ebb and flow which characterizes the sale of a liquid commodity produced in one place and used elsewhere—usually far away. In short, commercial inventories are operationally essential, while emergency reserves are not. This also explains why oil storage tanks are a natural (if exceedingly ugly) accoutrement of the oil trade everywhere, while emergency stocks would not exist if governments did not decide on their size and location, and who should pay for them.[63] (They are also—for security rather than aesthetic reasons—usually well hidden.)

Swedish oil stockpiling policies go back to the years immediately preceding the Second World War, when oil importers were for the first time enjoined by law to keep in storage, at their own expense, obligatory stocks of certain petroleum products equivalent to a percentage of the previous year's sales. These provisions continued to be in force until 1957, when the Riksdag, faced with the ever-increasing consumption of oil during the expansive postwar years, decided instead to pursue stockpiling policies in the more ambitious form of periodic programmes. Four such programmes have been in effect since then.

During the initial period, running from 1958 to 1962, a distinction was for the first time made between emergency stocks to be used during a war (*krigsreserven*), and those reserved for contingencies when the normal flow of oil is hampered by peacetime military or political crises beyond Sweden's borders (*försörjningsreserven*). While the extent of the former was kept secret, the size of the latter was defined not as a percentage of sales, as previously, but in terms of the volume consumed during a typical Swedish winter. It was also specified that the amounts of gasoline, kerosene and motor oil required for wartime purposes were to be kept in specially fortified storage facilities. For the first time, major petroleum consumers—and thus not only importers—were also required to keep emergency stocks. Three-fifths of the cost of the programme was to be covered by very favourable government loans (mainly to help finance the construction of new storage facilities), while the rest was to be borne by those required to keep stocks.

On the whole, the second programme (1963–9) followed the blueprint established by the first. The size of reserves was increased, however, and the name of the second type of stocks changed—to *avspärrningsreserven*—to emphasize its national security implications. Once again, however, no specific quantities of either type were publicly specified. The government continued to provide generous loans, while a consumer levy was for the first time introduced to cover all additional costs. This levy accrued to, and was administered by, a special fund run by the oil industry.

The third programme ran for eight years altogether (1970–7), at the end of which all government subsidies were terminated. New rules regarding additional types of oil products to be stocked were introduced, as were requirements about the geographic dispersal of war-related stocks and their storage facilities. It was also during this period that Parliament took a first step towards recognizing a third type of energy crisis situation, namely disruptions caused by supply factors as such—so-called *fredskriser*—rather than by strictly military or political developments, and as a consequence enacted a substantial extension of Sweden's existing stockpiles. In the wake of the first oil crisis further increases in existing inventories were introduced in both 1974 and 1975, the latter in response to the establishment of the IEP the previous year, to which Sweden was a signatory. After a commission report a new and integrated policy on emergency stockpiling was finally enacted by Parliament in 1977, incorporating the provisions of the IEP as well as those pertaining to the first two types of crisis situations sketched above. This programme has, with some significant modifications, covered the years 1978–86.[64] The more important details of this policy are the following.

The new energy reserves created by this programme are to be owned by the state and financed via a special consumer levy on energy (*beredskapsavgiften*). The specific goal of this programme is to fulfil IEA stockpiling obligations—ninety days of net imports—without having to use stocks reserved for other emergency purposes. They are also to be segregated in special storage facilities and to be directly controlled by the National Board for Economic Defence (ÖEF).[65] Full compliance with IEA obligations was, however, not envisaged until 1984 (rather than 1980), as stipulated in the IEP. Given the energy projections available in 1977, this implied that 9.6 million cubic metres of crude oil were to be acquired during the period in question, in addition to 1.8 million cubic metres of oil products. Of this total amount, 8.6 million cubic metres were to constitute the emergency reserves owned by the state (*fredsreserven*). By the beginning of 1980, about a quarter of this goal had been achieved.[66]

During the same year, a commission appointed by the government to provide a mid-term evaluation of the programme proposed the first of a series of changes which would occur during the rest of the period considered here.[67] As a result, part of the stocks kept for war purposes were now transferred to those reserved for the second type of crisis; and, for the first time, existing commercial reserves were henceforth to count as part of the emergency stockpiles reserved for the third rather than the second type of crisis.[68] Because of lower energy consumption projections and the effectiveness of conservation measures (and, as it should always be added in these contexts, the general economic downturn during the late 1970s), a revised and substantially lower goal for the state's reserves of crude oil was also set: 6.5 million cubic metres instead of the 9.6 million cubic metres previously stipulated. This figure did not represent any changes with respect either to fulfilling IEA obligations or maintaining other national goals in the event of supply disruptions. However, the time limit set for achieving this fill-up was extended from 1984 to 1986, mainly as a consequence of the sharp price increases following the second oil crisis.

For the first time, the report of this committee also contains a systematic and operationalized discussion of the crisis concepts which underlie the Swedish stockpiling programme.[69] *Wars* are defined as situations when Sweden is involved in warfare and when such actions are conducted on Swedish territory. A *blockade* exists when the country itself is not involved in warfare but when her trade relations are increasingly being disturbed as a result of such activities elsewhere in Europe. In both instances it is assumed that serious disturbances in the net supply of petroleum will occur. In the case of war, emergency planning is based on the assumption that, on the average for the projected period as a whole (the length of which is not specified), only 5 per cent of normal imports will be available. The comparable figure in the case of a blockade is put at 25 per cent (ranging from 90 per cent to 0 per cent during different phases of the period as a whole, stipulated as one year). Hence both of these crises will require very substantial efforts in restructuring the energy system and curtailing and redirecting the use of the various forms of energy available. Thus, for example, the assumption is that the average consumption of energy during a blockade will be 60 per cent of normal consumption, despite the availability of only a quarter of normal oil supplies. It is also stressed that since the one type of crisis does not exclude the other (e.g. a blockade can in due course evolve into regular warfare), separate contingency plans must continue to exist for both. Stocks reserved for the one type of emergency situation are not, in other words, to be used during the other type of crisis.

Peacetime crises occur when normal standards in the national welfare of the country—defined in terms of industrial production, employment, export and domestic consumption—cannot be sustained as a consequence of interruptions in the supply of one or more commodities crucial to maintaining these normal levels.[70] Such crises are distinguished from two other, somewhat similar, types of situations: adverse economic periods which are the result of scarcity, competition, structural changes or similar recurring (and hence 'normal') contingencies of international trade; and situations which arise when 'conflicts lead to actions selectively directed at Sweden for purposes of gaining a political advantage'.[71] The projected duration of this third type of crisis, and hence the size of the stockpile to be reserved for it, is based not on any specific Swedish calculations or crisis scenarios but simply on the IEA obligation of keeping a national stockpile equivalent to ninety days of net imports. In view of the rationale underlying the first two types of emergency reserves (to help preserve Swedish *national* autonomy), while the IEA commitment is an international undertaking premised on the necessity of maintaining a close link between such national policies and multinational oil allocation decisions during emergencies (a *supranational* principle), it is not surprising that the committee also reiterates the previous Swedish policy of not including the first two types of stocks in the calculation and implementation of her IEA obligations.

These changes were enacted by the Riksdag in 1981, when it was also decided that the programme should be re-evaluated a second time before the end of 1983. In 1982 the financial structure underlying the state's emergency holdings of crude oil was changed. Instead of the levy charged for these purposes and held in a special fund, an equivalent increase in the energy tax was introduced, and the emergency reserves financed via ordinary annual budgetary allocations to the National Board for Economic Defence. The proposed re-evaluation of the programme was conducted as planned, and in 1984 two important changes to the programme were made by the new Social Democratic government.

The most important change was the proposal to amalgamate the three types of reserves into one (referred to by a real tongue-twister: *försörjningsberedskapsreserven*).[72] According to the government bill, this did not mean that different regulations would not continue to be in force with respect to how stocks for the various types of crises were to be stored and used. However, it would give the government somewhat greater flexibility in the use of both private and public stockpiles in the event of a crisis. Nor would this change affect Sweden's international obligations, since the ninety day commitment would continue to remain fully in force. Second, the government proposed for the first time to decrease rather than increase the stocks under its jurisdiction. This decision to sell rather than to buy was based on new projections indicating a further substantial decline in the use of oil during the remaining period of the programme. Since a reserve of only 2.7 million cubic metres of crude oil would be needed to maintain the ambitions of the programme, the government was now in the happy position of being able to dispose commercially of 1.2 million cubic metres at its own pace and discretion.[73]

The Nuclear Issue

'Nuclear power was conceived of as the only indigenous source that would offer independence, be economical, and at the same time establish Sweden on the technological forefront'.[74] This, in essence, was the view that emerged in Swedish political circles during the mid-1950s, when—as we have seen above—imported oil was for the first time making substantial inroads into the Swedish energy system. By the beginning of the 1970s eleven nuclear reactors had been licensed by the government (of which four were already in operation), and the utilities were calling for the construction of thirteen more by 1990, in addition to planning actively for the introduction of fast breeder reactors.[75] None of the political parties was opposed to this development; indeed, all welcomed it, though for different reasons:

> The debates that did take place were over different technical aspects, such as the relative merits of LWR (light-water reactors) over HWR (heavy-water reactors), together with the interconnected issue of the role of the state versus private enterprise in what was regarded as a future growth industry. Nuclear power as such was never questioned; it was seen as the only way to avoid exploitation of the remaining hydropower potential in northern Sweden.[76]

And yet, by 1980 the tables had been almost completely turned, and the production of electricity in nuclear plants was regarded by all political parties to constitute a parenthesis in Swedish energy policy, despite a dependence on foreign oil at that time which had hardly ever been more pronounced. This, at least, was the result of a national referendum on 23 March that year, terminating half a decade of fierce political battle dominated almost completely by the nuclear energy issue.

How, in view of what has been noted above about widespread political perceptions of the need to decrease Sweden's heavy dependence on imported oil, is this rejection of an available alternative to oil to be explained? In order to answer this question it is necessary to provide a sketch of how, during the years in question, the nuclear issue moved from being characterized by widespread consent and optimism to becoming a major divisive factor cutting across traditional political dimensions and alignments.

In this connection it is important to note, first of all, that prior to 1973 Sweden did not possess a national policy on the role of nuclear power. This conforms to an underlying theme of broader Swedish industrial policy, namely that 'the State creates the environment conducive to change and pays some of the cost but leaves it to others to take specific decisions'.[77] This had left the energy producers wide scope for expansion, and hence it is not surprising that all plans up to this point had been formulated by the electric utilities themselves and co-ordinated by the Central Operating Management (*Centrala driftsledningen*), the organization representing electric companies using the national grid. The government's role had lain mainly in the licensing of nuclear plants, while Parliament had been involved only in the financial aspects of nuclear power investments made by the country's publicly owned utility, the State Power Board.

However, a conjunction of two distinct developments put an effective end to the freedom of action enjoyed by the Swedish energy industry. We have already discussed the first of these: the oil crisis in 1973–4, which, after a period of shortages and an intensive public discussion, led to the establishment of Sweden's first comprehensive energy policy in May 1975. Obviously, in such a policy the role of nuclear power was destined to constitute a central component, given its purported viability as a substitute for both imported oil and hydroelectric power. The second development, the origins of which slightly antedate the oil crisis, was the emergence, almost overnight, of a political movement questioning the feasibility of the nuclear option.

The first intimations of this incipient public reaction were contained in a query raised in the Riksdag in the autumn of 1972 by a newly elected provincial member of the Centre Party regarding the normative implications of nuclear energy.[78] This seemingly peripheral event, constituting one of the more oft-cited additions to recent Swedish political folklore (on the theme of the mouse that roared), led to a broader debate in early 1973 on reactor safety and waste management. It also signalled the beginning of the Centre Party's increasingly active—although as yet not vociferous—stand against nuclear power in general. This split between the non-socialist parties was quickly followed by a similar split within the socialist bloc, when the Communists—somewhat surprisingly—also decided to question the feasibility of the nuclear option. The result was the acceptance of a resolution by Parliament in May 1973, stating that 'no decisions in favour of the further expansion of nuclear power should be taken until new basic material, including information on the state of research and on development tendencies has been put before the Riksdag'.[79] The most important consequence of this resolution was the creation of study commissions on reactor safety and on waste handling and—during its deliberations, which were to be concluded the following year—the decision to institute a moratorium on the expansion of nuclear power beyond the eleven in service or under construction at that time. Equally importantly, by setting into motion these developments, Parliament in practice also 'requested a national plan for nuclear power and . . ., in effect, took over from the government ultimate responsibility for nuclear power'.[80] In other words, nuclear energy was now for the first time declared controversial and placed squarely on the political agenda, where it remained and played an increasingly pivotal role during the rest of the decade. Also, as a result of the oil embargo and the rise in the price of oil, it became clear that any decision on nuclear power had to be extended to a decision on future Swedish energy policy in general. It was at this point that the nuclear issue became inextricably linked to the vicissitudes of the international oil crisis and hence to broader issues of industrial policy, economic development, employment, environmental concern and, consequently, to Swedish national welfare as a whole.

The Social Democratic government tried to defuse the political problems associated with the nuclear question by pursuing a vigorous energy policy based primarily on conservation measures. It certainly did not wish to politicize the energy policy arena; but at the same time public pressure and the party's precarious electoral position made a response to the emerging challenge to its previous commitments to a high rate of energy growth and to the development of nuclear power unavoidable. This response is contained in the comprehensive energy policy discussed above, which, while it contained plans to extend the number of nuclear plants to thirteen (from eleven), at the same time tried to downplay this fact by concentrating on the need, as an overarching policy, to decrease the growth rate of energy demand. Sahr has argued that this pressure on the government to revise its energy policies also goes a long way to explaining why Sweden decided to pursue forceful conservation policies when most other countries were still showing clear indications of suffering from the so-called energy syndrome. He writes as follows:

> The argument has been that serious energy conservation efforts were accepted in Sweden early relative to other nations largely because conservation promised to help resolve both a substantive issue—dependence on imported oil—and a set of political problems. The political problems for the Social Democrats arose primarily over nuclear power, both within the electorate, including among Social Democrats, and within the energy bureaucracy and industry. Conservation could help both by reducing the need for additional energy sources, including both nuclear power and imported oil, and thereby slowing growth of these politically difficult energy sources, and by serving as a point to emphasis to deflect attention from nuclear power.[81]

However, if this interpretation is correct, the strategy of the government was less than fully successful.

First of all, although the government tried hard, it was difficult to change prevalent and deeply ingrained—and internationally pervasive—beliefs within the industrial and economic sectors about a positive link between energy growth and economic expansion. More important and immediately troubling to the government, however, was the fact that this policy did not succeed for long in preventing political parties within both blocs from raising the government's nuclear policy as a partisan issue. Since the election in 1973 had created a precarious Parliamentary balance of 175–175 between the socialists and the non-socialist parties, this development gave the incumbent Social Democrats occasion for becoming increasingly concerned in the face of the upcoming 1976 elections.[82] Such fears also turned out to be well-founded, particularly since, as Vedung has pointed out, an issue which could not be classified in terms of a left–right dimension—namely nuclear energy—for the first time came to dominate a Swedish election. This could obviously not benefit an ideologically highly charged and homogeneous party such as the Social Democrats, especially not when it itself was becoming increasingly divided in regard to energy policy. Thus, although the government tried valiantly to shift attention away from the nuclear issue to energy policy in general, and especially to the need for conservation in the face of the country's huge dependence on foreign oil, public discussion continued to focus on the simpler and more palpable issue of how many nuclear plants each party favoured. Olof Palme's only consolation was that this question was as much a problem for the unity of the non-socialist bloc—a unity which for all practical purposes was a prerequisite for a change of government—as it was for the socialist parties. Nevertheless, the result of the election was that for the first time in forty-four years the hegemony of the Social Democratic party was broken; and it is generally acknowledged by political friends and foe alike that the issue of nuclear energy played an important, and perhaps a crucial, role in this historic outcome.[83]

This issue was also instrumental in destabilizing the political arena itself during this election and for a long time to come. Vedung's analysis of these divisive changes is highly pertinent here.[84] He has suggested that as a consequence of the introduction of the energy issue to Swedish politics in the 1970s, the traditional and highly pervasive left–right dimension in politics—placing the Social Democrats and the Communists on the left and the three non-socialist parties on the right—was overlaid by a fundamentally new dimension based on an ecology-growth scale. This latter polarity placed the Social Democrats, the Liberals and the Conservatives on the 'growth' end and the two other parties on the 'ecology' side of this new dimension. Consequently, while the Social Democrats had usually been in agreement with the Communists on traditional 'class' issues (and had often been dependent on its parliamentary support while in government), this did not hold with regard to the set of issues associated with the energy debate. More importantly, in view of the result of the 1976 elections, the non-socialists—who during most of their joint history had aligned themselves on the same side with regard to 'class'-oriented policies—were now faced with a situation in which the Centre Party sharply diverged from the other two (and was in agreement with the Communists) on the main issue facing the new government.

Not surprisingly, this turned out to be an untenable position in the long run. During the election Thorbjörn Fälldin, the Centre Party leader (and premier designate in the event of a non-socialist victory), had promised to dismantle nuclear energy altogether by 1985, insisting that stopping 'the march into the nuclear power society' was an uncompromisable commitment and his party's prime objective once in power. He had also made categorical commitments not to participate in a government permitting the fuelling of new plants. However, even before the new government was properly constituted, Fälldin—under threat of having to form a minority government—had to make a first compromise by allowing a newly completed reactor on Sweden's west coast, Barsebäck 2, to be activated. Since this accommodation was announced in the very first official statement issued by the new government, it immediately led to talk about the new prime minister's trustworthiness once installed in office. (Olof Palme, of course, was furious.) Being by most accounts a self-consciously honourable man, such intimations of having 'given in' or 'sold out' on nuclear power made any further compromises in this area virtually impossible. Consequently it came as no surprise that 'attempts to reconcile the irreconcilable resulted in what one . . . participant called a "crisis each month" on energy until the government fell over the nuclear issue two years after taking office'.[85] In other words, the first Fälldin government (there were to be two more) decided—or was forced—to leave office on the very same issue on which it had come to office.

The specific problem which led to this final crisis involved the question which all along had been the Centre Party's main concern during this period, namely the safe disposal of spent nuclear fuel. (This was before it was known that nuclear reactors could, in practice, also be surprisingly accident prone.) The decision to allow Barsebäck 2 to be fuelled had been accompanied by a proviso that this approval would be subject to the signing, not later than in October 1977, of a satisfactory agreement between the government and the owner of that plant on the reprocessing of its spent fuel.[86] (Reputedly, Fälldin had got this idea from a referendum in California involving the terms governing permission to fuel new reactors there.)[87] Subsequently, in the Nuclear Power Stipulation Act (*villkorslagen*), approved by Parliament in April 1977, this requirement was extended to cover all nuclear plants not licensed prior to 8 October 1976.

More specifically, this act stipulated that before being allowed to fire a plant, its operator must either:

1. present a contract which adequately provides for the reprocessing of spent fuel, and also show how and where the final deposition of the highly radioactive waste resulting from reprocessing can be effected with absolute safety, or
2. show how and where the spent but not reprocessed nuclear fuel can be finally stored with absolute safety.[88]

As various commentators have pointed out, the motives underlying the Stipulation Act varied considerably between the main protagonists of the nuclear energy issue. Sahr, for example, has commented on these as follows:

The bill was central to the attempt of the Centre party to stop the development of nuclear power in Sweden. Yet the bill did not finally decide any element of energy policy but only postponed it. In addition, it proved to be a particularly 'blunt instrument' in relation to guiding energy policy decisions, in large part because—as part of the compromise process—such large ambiguities were built into the wording of the law. What was 'adequately' to mean? And how was 'has shown' to be interpreted? Centre party leaders appear to have believed that it was impossible to satisfy these requirements within the time period given, so that they would result in Sweden's discontinuing the use of nuclear power. Leaders of the other parties did not share this view.[89]

Lönnroth and his colleagues present a similar interpretation when they note that the

the law was seen by the pronuclear parties as a way to force the utilities to put forward a scheme for waste disposal, and the parties had no serious doubts that the government would be able to accept the proposal rather soon and thus grant the licenses. The antinuclear party, however, saw the law and its safety conditions as an unsurmountable barrier to the further expansion of nuclear energy.[90]

In order to give the new government time, a special Energy Commission was also created, endowed with the task of proposing, by the spring of 1978, a broad solution to the problems surrounding nuclear power.

The electric utilities and the nuclear industry immediately set into operation technical feasibility projects in response to the requirements of the Stipulation Act, and in September 1977, the owners of Barsebäck 2 petitioned for approval of its continued operation on the basis of the results of one of these studies—the so-called Nuclear Fuel Safety Project—and a contract with a French nuclear reprocessing firm. Although the Centre Party, in line with its a priori view regarding the practical infeasibility of fulfilling the requirements of the Act, maintained that the specified conditions had not been met, the other two government parties successfully argued that the signing of the contract with Cogema, the French firm, constituted an acceptable solution as far as this particular reactor was concerned (for reasons indicated above, Barsebäck 2 was considered a special case).

In early 1978 similar arguments were presented by the nuclear industry for the next two reactors on line. Once again, Centre Party leaders fought approval on the grounds of the inadequacy of the proposals submitted to the government. Although the response was not affirmative when a compromise was finally made public on 29 September 1978, this result was nevertheless regarded as a capitulation by the Centre Party, since the government simply postponed taking a definite stand by requiring further test drillings of suitable Swedish bedrock sites for the final deposition of nuclear waste. The government also designated the Swedish Nuclear Power Inspectorate (SKI, *Statens kärnkraftsinspektion*) as ultimate arbiter of the acceptability of such sites, which in effect—contrary to the interests if not intent of the antinuclear forces—depoliticized the issue by transferring into the hands of technical experts and a public agency. Lönnroth has thus argued that

it seems that the Centre party fell into its own trap—once a technical scheme is suggested and subject to a certain amount of scientific screening the burden of proof tends to shift. It became as difficult to disprove the scheme (which the coalition partners demanded of the Centre party) as it is to prove the safety beyond doubt (which the Centre party required). . . . Thus there was no way through which an agreement could be reached on how to interpret the term 'absolute safety' and where to place the burden of proof.[91]

Indeed, a consequence of this settlement was that permission to fuel the two reactors could in practice be forthcoming within a few months, particularly since the Nuclear Power Inspectorate had already previously found the Nuclear Fuel Safety model satisfactory.[92]

At this point—facing fierce attacks on his credibility in the news media, by the environmentalists and within the ranks of his own party—the prime minister tried to resist being further tarred with the nuclear brush by suddenly introducing new conditions regarding the future acceptability of nuclear energy. More specifically, Fälldin now insisted, in the form of an ultimatum, that the construction of Forsmark 3 (Sweden's eleventh plant) be discontinued altogether, or in the event that this was not acceptable, that an advisory referendum be held to decide on the future of nuclear power in Sweden. Since the first would effectively kill the nuclear industry, and the second possibly squash the future viability of the nuclear option as well, neither of these demands was acceptable to his government's two junior partners. After an unprecedented media blitz lasting for almost a week, the demise of the first Fälldin government was announced on 5 October 1978.[93]

A new government was formed by Ola Ullsten and the Liberal Party, based on thirty-nine of the Riksdag's 349 seats. In a bill on energy published in early March 1979, it called for the operation of twelve reactors—but no more—by 1990. However, as we have already noted above, before a decision on this programme could be made, the accident on Three Mile Island radically changed the whole situation. Olof Palme, in one of the more notable turnabouts in recent Swedish history, now acknowledged his party's willingness to participate in a referendum on the future of nuclear energy. The Liberals and Conservatives quickly followed suit; they also agreed with the Social Democrats that the referendum should be held after the 1979 elections. Parliament set a date for 23 March 1980.

All three referendum 'lines' involved putting an end to the future expansion of nuclear energy, that is closing the so-called 'nuclear parenthesis' in Sweden. The question was how quickly this should be done, and therefore what to do about the reactors in service, under construction or still on the drawing boards. The first two referendum options recommended, in verbatim terms, the construction of at most twelve reactors, after which nuclear energy 'be phased out at the rate that is possible with regard to the need for electric power for the maintenance of employment and our welfare. . . . Safety considerations are to determine the order in which the reactors are taken out of operation (service)'.[94] The third option—regarded as the 'no' choice—was much more specific and emphatic in its demand for the 'phasing out of the six reactors now in operation within a period of, at most, ten years'.

The difference between the first two 'lines'—associated with, respectively, the Conservatives and the Liberals/Social Democrats—had mainly to do with the ownership of the nuclear industry (line 1 against public ownership, line 2 for), and nothing at all with the future utilization of nuclear power as such. In this sense both were regarded as generally affirmative as against the negative stance of the third option (recommended by the Centre and Communist Parties). However, it is generally agreed that the proponents of line 1 were more positive to nuclear energy than those of line 2, whose viewpoint came to be regarded as something of a compromise option.

After an unusually acrimonious and long campaign, the two 'yes' options received 58.0 per cent of the votes cast as against 38.7 per cent for line 3. Line 2 was regarded as the winner, since it by itself received more than line 3 (39.1 per cent). In short, those generally more positive to nuclear energy had won the referendum by a clear margin. At the same time, however, the referendum also implied a Pyrrhic victory for these groups, since it gave a clear indication that nuclear energy would henceforth be regarded as an interim feature rather than a permanent component of the Swedish energy system. In concrete terms this meant that as a result of the referendum all nuclear plants in Sweden would have to be put out of commission at the latest by the year 2010. All five political parties accepted this result of the referendum.

The referendum also accomplished something else which is no less significant in the present context: almost overnight it removed the energy issue from the political agenda. Interviews conducted within two months of the referendum thus showed a significant drop of interest in the nuclear issue. In this connection Sahr notes that 'all interviewed participants, from all positions, indicated that the long battle over nuclear power, extending back over more than five years, had taken its toll in "battle fatigue" and that they were simply glad it was over'.[95] Since the debates preceding the referendum 'touched almost every Swedish voter and cut across family, political and social alliance', this was perhaps just as well for all concerned.[96] It was also a reaction which was to last beyond at least the 1982 general election, in which the nuclear issue played a minor role, both in the news media and in public debate.[97] Since then the practical implications of the referendum have continued to be discussed on and off, but hardly the nuclear question itself. This situation did not change until the accident in Chernobyl in the spring of 1986.

The 1979 general election preceding the referendum once again led to a non-socialist government headed by Thorbjörn Fälldin. Its first energy proposal, submitted to Parliament shortly after the referendum, reaffirmed the result of the latter, namely that nuclear energy would be phased out over a thirty-year period. The government also approved the fuelling of the four reactors completed subsequent to Fälldin's resignation in 1978. In early 1981 the Riksdag was presented with a more comprehensive energy programme, the first to revise the goals of the 1975 decision.[98] It *inter alia* contained a more detailed policy for phasing out nuclear power, a process which was to be set into motion by 1997 and to end no later than 2010.[99] Although the second Fälldin government was dissolved over a disagreement with the Conservatives on tax policy, and the third—and, as it turned out, final—Fälldin administration was formed in early May that year (together with the Liberal Party), this had no effect on this bill, which was passed by Parliament later that spring. Thus—and here I gratefully if irreverently revert to quoting poetry after a long stretch of prose—ends the story of nuclear energy during the period considered here: not with a bang but a whimper.

CONCLUSION

The above has been a description of the main thrust of Swedish energy policy during the decade with which we are here concerned. The following conclusions emerge from this analysis. (1) Sweden was undoubtedly successful in counteracting the so-called 'energy syndrome', particularly with respect to achieving credible energy conservation programmes and a decreased reliance on imported and especially Middle East oil for her energy needs. Thus, for example, compared to the United States, Sweden at present uses about half the energy per dollar of gross domestic production.[100] (2) Sweden set into motion an extraordinary R & D programme in the area of energy policy, albeit that its effects on alternative, domestically produced energy sources have as yet on the whole remained negligible. However, as surmised in a recent study, the future adoption of new energy-efficient technology—not necessarily more capital-intensive than the present technology—could economically lead to the reduction of energy demand by half, even under conditions of economic growth.[101] (3) Sweden did follow IEA requirements regarding oil stockpiling policies, though certainly not as quickly as required by the IEP. She has also been a keen participant in the other programmes of the IEA, integrating these with the country's own emergency planning measures. This reflects a long-standing tradition of working closely with functional international organizations of this kind. (4) Although extensive plans existed in the early 1970s for making Sweden one of the most nuclear-energy-oriented states in the world, this trend was reversed as a consequence of a long and unusually acrimonious political battle, ranging over both political blocs and fuelling partisan strife for almost the whole decade.

Given these developments, and especially the obvious centrality of the energy issue on the political agenda during the decade in question, must one not conclude that, on the whole, Swedish energy policies have been responsive to the challenges posed to her national welfare and security as a consequence of her dependence on the import of foreign sources of energy? This is obviously an answer which lies close at hand, and it is one which most Swedish policy analysts and decision-makers would probably give. However, in my view it is not possible to face this crucial query head-on before we have dealt with an additional policy domain which is central to all issues concerning the country's international interdependence: Sweden's policy of neutrality in foreign affairs. Indeed, it is only in direct relation to the latter that we can in fact tackle the main issue raised in this analysis as a whole, namely the relationship between Sweden's energy vulnerability and her national security. Hence, before trying in the final chapter of this study to respond to the question reiterated above, we must perforce first consider the special logic on which Sweden's neutrality is based—the logic which constitutes the linchpin around which Sweden's national sovereignty and autonomy revolves.

NOTES

1. James L. Plummer (ed.), *Energy Vulnerability* (Cambridge, Mass., 1982), p. 327.
2. One has, however—my esteemed colleague Evert Vedung. On the other hand, he is *extremely* knowledgeable about such policies.
3. For the most exhaustive analysis of this programme, and at the same time an intriguing policy analysis in its own right, see Björn Wittrock and Stefan Lindström, *De stora programmens tid—forskning och energi i svensk politik* (Stockholm, 1984).
4. A good introduction to this state of affairs is provided by Jan-Erik Svensson, Elsa Hakkarainen and Anders Eriksson, *Kunskap om energi*: *En inventering av svenska energisystemstudier och några hypoteser om forskningsbehov* (Stockholm, 1982). Not a single one of the over 150 studies on which they report pertains to the foreign/security policy area. A similar dearth is exemplified in Evert Vedung's evaluation of Swedish energy projects, entitled *Energipolitiska utvärderingar 1973–81* (Stockholm, 1982). In English, the two best-known books are by the same three authors: Måns Lönnroth, Thomas B. Johansson and Peter Steen, *Solar versus Nuclear*: *Choosing Energy Futures* (Oxford, 1980) and *Energy in Transition* (Berkeley, 1980). Both are reports prepared for the Swedish Secretariat for Futures Studies and contain little more than general reflections on the link between energy and foreign/security policies. They are also somewhat dated, pertaining to conditions prevailing mainly during the mid-1970s.
5. See Wittrock and Lindström, pp. 21, 42ff, as well as the former's 'Beyond Organizational Design: Contextuality and the Political Theory of Public Policy', in Kenneth Hanf and Theo A.J. Toonen (eds.), *Policy Implementation in Federal and Unitary Systems* (Dordrecht, 1985), pp. 22f. Also see L. Lundgren, *Energipolitik i Sverige 1890–1975* (Stockholm, 1978).
6. Quoted in Wittrock and Lindström, p. 42. Author's translation.
7. *SOU 1956:46 Bränsleförsörjning i atomåldern*, p. 26. Author's translation. These types of study commissions are unique to Swedish policy making, and are appointed by the government, acting through its ministries, and usually range in size from one to fifteen or twenty members, assisted by expert groups. They are based on directives issued by the government, and are published in the SOU series, *Statens offentliga utredningar*, i.e., literally, the state's official investigations. Before these reports are used by the government, they are circulated by means of the *remiss* procedure, when all interested parties are invited to respond

to them. For a succinct description pertinent to the present study, see Robert. C. Sahr, *The Politics of Energy Policy in Sweden* (Ann Arbor, 1985), pp. 10f. When these reports are cited more than once below, they will in each such subsequent case be abbreviated in the following form (this example is pertinent to the above source): *SOU 1956:46*.

8. See, in particular, K-G Ljungdahl's report, published as *Bränsle och Kraft* (*SOU 1951:32*) and *SOU 1956:46*. The discussion in Wittrock and Lindström, pp. 61ff, is indispensable here. The cited epithets are to be found in Wittrock, 'Beyond Organizational Design', p. 22.
9. Here the pertinent official document is *SOU 1956:11 Atomenergien*, which is the hastily prepared report of the Royal Commission on Atomic Energy, constituted the previous year. See also Sahr, pp. 74, 207.
10. Lönnroth et al., *Energy in Transition*, p. 6.
11. See, for example, Nigel Lucas, *Western European Energy Policies: A Comparative Study* (Oxford, 1982), pp. 110ff.
12. This section is largely based on Lönnroth et al., *Energy in Transition*, pp. 6f, and Wittrock and Lindström, pp. 92ff.
13. Lönnroth et al., *Energy in Transition*, p. 7.
14. See Sahr, pp. 15f.
15. Leon N. Lindberg, 'The Energy Crisis: A Political Economy Perspective', in Leon N. Lindberg (ed.), *The Energy Syndrome: Comparing National Responses to the Energy Crisis* (Lexington, 1977), p. 370.
16. Måns Lönnroth, 'Swedish Energy Policy: Technology in the Political Process', in Lindberg, p. 258.
17. See Lindberg.
18. Lindberg, p. xi.
19. See Sahr, pp. 3ff.
20. For the text of this bill, see *Regeringens proposition 1975:30 Energihushållning m.m.*, pp. 4ff, and especially pp. 11ff. (All government bills cited in the text more than once will in each subsequent case be abbreviated in the following way, in this particular case relevant to the bill cited above: *Prop. 1975:30*.) See also the discussion by Lönnroth, pp. 255f, and Sahr's description of this decision in Sahr, p. 4.
21. Sahr, p. 4.
22. Lönnroth, p. 263. This whole paragraph depends extensively on Lönnroth's discussion in ibid., pp. 263ff.
23. See Lönnroth, pp. 263ff.
24. Sahr, p. 52.
25. See Theodore S. Lowi, 'American Business, Public Policy, Case Studies and Political Theory', *World Politics*, vol. 16 (1964), pp. 677–715, and 'Four Systems of Policy, Politics, and Choice', *Public Administrative Quarterly*, vol. 32 (1972), pp. 298–310. See also my own discussion in 'Foreign Policy and the Democratic Process' *Scandinavian Political Studies*, vol. 4 (1981), pp. 92ff, and—more pertinent here—Sahr, pp. 55ff.
26. Sahr, p. 56.
27. Sahr, p. 62.
28. For a discussion of how the 2 per cent figure was arrived at, see Appendix 1 in Sahr, pp. 185ff.
29. Evert Vedung, 'Main Trends in Swedish Energy Policy in Response to the Oil

Crisis', paper presented to a workshop of the European Consortium for Political Research in Salzburg, 13–18 April, 1984, p. 6. The bulk of this paper is a revised version of the appendix to Vedung.

30. Vedung, p. 8.
31. In addition to the source quoted immediately below Table 3. 1, see also Vedung, p. 10, and Sahr, p. 49.
32. Vedung, p. 12, and *Prop. 1975:30*, Appendix 1, p. 360.
33. In addition to the source quoted immediately below Table 3. 2, see also Vedung, p. 13, and Sahr, p. 49
34. *Regeringens Proposition 1977/78:76 Med energisparplan för befintlig bebyggelse*, pp. 1 and 28f gives a figure between 32 and 39 TWh, but Parliament specified the more exact figure indicated above. See Vedung, p. 14.
35. These are total/net figures, that is after considering process losses.
36. See *Regeringens proposition 1978/79:115 Om riktlinjer for energipolitiken*, pp. 2, 115ff.
37. See Vedung, p. 15.
38. See *Prop. 1978/79:115*, p. 1.
39. *Prop. 1980/81:90*, p. 92.
40. See *Prop. 1980/81*, *Bilagsdel*, p. 15, for an illustrative diagram.
41. See *Regeringens Proposition 1984/85:120 Riktlinjer för energipolitiken*.
42. *Prop. 1984/85:120*, pp. 43f.
43. See *Prop. 1975:30*, Appendix 1, pp. 124ff.
44. See Wittrock and Lindström, pp. 117ff.
45. Björn Wittrock, Stefan Lindström and Kent Zetterberg, 'Implementation Beyond Hierarchy: Swedish Energy Research Policy', Stockholm University, mimeo, May 1981, p. 4.
46. Ibid., p. 1.
47. Ibid., p. 10.
48. See ibid., pp. 10ff.
49. Wittrock and Lindström, my translation.
50. For an official description, see for example *Energy Research and Development in Sweden 1978/81*, DFE Report No. 13 (Stockholm, 1978), pp. 24ff. This, and all other numbered reports issued by the *Energy Research and Development Commission* will in future be referred to only in terms of its number in the DFE series.
51. DFE Report No. 13 (1978), p. 25.
52. Wittrock, Lindström and Zetterberg, p. 6.
53. Ibid., p. 16.
54. Ibid.
55. *SOU 1974:72 Energiforskning: Program för forskning och utveckling*, p. 9.
56. DFE Report No. 13 (1978), p. 9.
57. DFE Report No. 42 (1982), p. 6.
58. See Wittrock, Lindström and Zetterberg, pp. 12ff, and DFE Report No. 49 (1982), pp. 11ff.
59. Wittrock, Lindström and Zetterberg, pp. 13f., give these figures, but do not provide any references.
60. See *SOU 1977:56 Energi—Program för forskning, utveckling och demonstration (EFUD)*.
61. See DFE Report No. 42, pp. 25f.
62. Wittrock, 'Beyond Organizational Design', p. 25.

63. See Edward N. Krapels, *Oil Crisis Management: Strategic Stockpiling for International Security* (Baltimore, 1980), pp. 56ff.
64. This short historical résumé is based mainly on *Ds I 1984:13 Energiberedskap för Kristid*, Appendix 2 (*Översyn olja 83*), pp. 15ff; and *Ds I 1984:10 Olja och kol: Förslag till lag om beredskapslagring*, pp. 18ff.
65. See *SOU 1980:41 Olja för kristid*, p. 25.
66. See *SOU 1980:41*, p. 63; and *SOU 1986:42 Bränsle- och drivmedelsberedskapen under kriser och i krig*, p. 29.
67. *SOU 1980:41*.
68. See *Ds I 1984:13*, Appendix 2, p. 34.
69. Although this discussion is not contained in the government bill which Parliament enacted in 1981, it must nevertheless be considered to reflect the 'official' viewpoint in the matter.
70. *SOU 1980:41*, pp. 62f.
71. Ibid., p. 63. My translation.
72. *Proposition 1983/84:110 Om vissa oljefrågor*, pp. 73f.
73. Ibid., p. 70.
74. Lönnroth et al., *Energy in Transition*, p. 137.
75. Ibid., pp. 138f.
76. Ibid., p. 139.
77. Nigel, p. 105.
78. For a verbatim report of this innocuous query, see Evert Vedung's *Kärnkraften och regeringen Fälldins fall* (Stockholm, 1979), p. 14.
79. Quoted and translated by Sahr, p. 21. See also Evert Vedung, 'Kärnkraften och partisystemet i Sverige', *Politica*, vol. 12 (1980), p. 118. This article contains an excellent analysis of the development of party orientations toward nuclear power during the 1970s; see pp. 108ff.
80. Lönnroth et al., *Energy in Transition*, p. 140.
81. Sahr, p. 73.
82. The constitutional basis for making possible this type of deadlocked parliamentary situation was subsequently changed; the Riksdag has 349 rather than 350 members at present.
83. See, for example, Sören Holmberg and Kent Asp, *Kampen om kärnkraften* (Stockholm, 1984), pp. 11ff.
84. Vedung, 'Kärnkraften och partisystemet i Sverige', pp. 109ff. See also Sahr, pp. 86f.
85. Sahr, p. 87.
86. See ibid., pp. 86ff.
87. In a recent and painstakingly researched analysis of these events, Sven-Erik Larsson (formerly an editor-in-chief of *Dagens Nyheter*, Sweden's premier morning newspaper) also maintains that the 'idea of laying down terms for the fuelling of new reactors . . . was an ace which he [Fälldin] kept up his sleeve and which he did not produce until the government negotiations following the election'. In other words, while the 'conditions governing fuelling were portrayed as a compromise solution', Fälldin's plan was to use this 'compromise' as an 'instrument which he—through a strict interpretation of the requirement—intended to exploit in rolling up the nuclear power front'. Sven-Erik Larsson, *Regera i koalition* (Stockholm 1986), p. 437. My translation. The reference to California is contained on p. 436.

88. *Prop. 1976/77:53*, p. 2. The translation quoted here is that cited by Sahr, *op. cit.*, p. 90. His account (on pp. 86–99) of the energy policies of the first Fälldin government is the best available in English, and therefore serves as a baseline for my own summary here.
89. Sahr, p. 90.
90. Lönnroth et al., *Energy in Transition*, p. 142.
91. Lönnroth, as quoted by Sahr, p. 96.
92. See Larsson, p. 440.
93. See Sahr, pp. 95ff, and, for two more detailed accounts, Vedung, *Kärnkraften och regeringen Fälldins fall*, pp. 42 ff and Larsson, pp. 365ff, 439ff (the latter in the form of an English summary).
94. Sahr, in an Appendix, gives us translations of all three referendum 'lines'. See pp. 189f.
95. Ibid.
96. Lucas, p. 123.
97. See Holmberg and Asp, p. 14.
98. See also above, p. 54.
99. See Sahr, p. 125.
100. See Thomas B. Johansson et al., 'Sweden Beyond Oil: The Efficient Use of Energy', *Science*, vol. 219 (1983), p. 360.
101. See the rosy picture painted in Johansson et al., pp. 355ff.

4 Foreign Policy, National Security and the Logic of Swedish Neutrality

Some years ago, in a sly play on Mies van der Rohe's celebrated maxim 'Less is more', a brash American architect launched the counter-slogan—heralding the first stirrings of the post-modernist turn in aesthetic diction—that 'Less is a bore'.[1] At the considerable price of being charged with levity, I would like to propose that these two epigrams not only pinpoint a major break in contemporary style and sensibility, but also suggest two of the most prominent characteristics of the doctrine of Swedish neutrality. Thus, although streamlined and functional to the core, this cornerstone of Swedish foreign policy has at the same time evoked little scholarly debate, intellectual excitement or, for that matter, domestic political strife. Instead, like the best of *Bauhaus*, it has tended to elicit grudging respect and widespread lip-service rather than emulation or enthusiasm. Perhaps for the very same reasons it has also been notably enduring—and greatly beneficial to the nation—as a principle of international concord.

The above is, however, no longer an adequate characterization of either the state of Swedish foreign relations or of its public discussion and scholarly analysis. Or rather, while the doctrine of neutrality itself as a principle has remained both unchanged and unchallenged, the present-day response to it differs significantly from the one suggested above. This is especially true for the period following the traumatic discovery in October 1981, on the rocky islet Torumskär, of a dilapidated Soviet submarine stranded in the middle of the best guarded naval passageway of Sweden's south-eastern archipelago. This event, and the series of submarine sightings which have been made intermittently since then, not only alerted international attention to the fundamental changes in the general strategic situation which have occurred around Europe's northern flank, but also for the first time in many years set into motion a domestic debate on the very foundations of Swedish national security. Faced with these incursions and Sweden's apparent inability in preventing further violations of her territorial integrity, it became increasingly difficult to derive comfort from either the political consensus or the ritual and pedestrian repetition of commonly accepted phrases which for so long had characterized domestic Swedish discussions—both public and scholarly—of national defence policies. Indeed, it has been argued that the effect of this so-called 'submarine crisis' was nothing less than the destruction of a world-view which had not only been taken for granted by the public at large for more than a generation but which had also dominated Swedish policy-making since the early 1960s. What came as a shock was the apparent willingness of at least one superpower to jeopardize this stability by repeatedly violating the territorial sovereignty of a neutral neighbour.[2]

Wilhelm Agrell, the most active participant in this renaissance of Swedish foreign-policy debate, has suggested that this erosion of confidence is a consequence of a fundamental misfit between the premises of an abstract doctrine—based on 'theoretical argumentation rather than practical experience'—and concrete reality; and that it is the sudden revelation of this misfit, rather than Soviet behaviour as such, which has so dramatically undermined 'the feeling of security which had dominated Swedish politics, debate and opinion' during the preceding decades.[3] Indeed, given the character of some of the altercations on this issue, one is perhaps even tempted to go one step further and to lend at least half an ear to the following pessimistic and undoubtedly censurable paraphrase of a memorable insight from Alexander Zinoviev's discussion of Soviet institutions: Swedish neutrality does not any longer embody a solution to a problem; it is merely the result of the search for a solution.[4] Whatever their validity, claims and insinuations of this kind obviously raise broader and long-term issues regarding the congruence between the assumptions of Swedish neutrality and the contingencies of the present-day international situation in Northern Europe. This, in turn, calls for an analysis not only of this doctrine itself but also—and perhaps more fundamentally—of *how* such an examination is to be tackled adequately in the first place. In what follows, I shall attempt to provide a conceptual outline of such a framework of analysis and evaluation.[5]

NEUTRALITY AND SWEDISH FOREIGN POLICY

The contemporary Swedish policy of neutrality, I would like to suggest, is based on that logic of action which has received its ancient and most enduring formulation in Ulysses' attempt to pre-empt the fatal seductiveness of the Sirens' song by issuing the following peremptory command to his hapless crew: 'you must bind me hard and fast, so that I cannot stir from the spot where you will stand me . . . and if I beg you to release me, you must tighten and add to my bonds'.[6] Jon Elster has referred to the type of behaviour exemplified here in terms of the logic of precommitment, the essence of which consists of the strategy of *a priori* excluding the availability of certain options (usually attractive) in order to achieve a desired future state of affairs.[7] Binding oneself—or forcing oneself to be bound—in this manner typifies the kind of imperfect but self-enlightened rationality which results from being weak and knowing it, and using this insight to set into motion an external, causal process aimed at limiting the adverse effects on oneself of such a condition. Higher-level preferences are made to override lower-level preferences, thus illustrating (in Elster's view) the innate and unique capacity of human beings to think not only of their immediate gain but also of the longer-term benefits which depend exclusively on the efficacy of indirect strategies.[8]

However, Elster has also noted that 'preferences are always relative to a past history of choices', choices which, in the case of Swedish neutrality, have had a profound and almost existential effect on modern Swedish politics itself.[9] Hence, before elaborating on the applicability of the above imagery and logic to the Swedish case, a short historical overview will first be provided. This brief sketch will show that contrary to popular belief (and, at times, the officially purveyed image of Swedish foreign affairs as well), the policy of neutrality has led a surprisingly checkered life during its long time-span. Indeed, it will be argued that the neutrality practised at present is essentially a post Second World War phenomenon premised on the existence of two superpowers, and hence that it differs fundamentally from the preceding forms of Swedish neutrality stretching back to the Congress of Vienna.

The Historical Roots of Swedish Neutrality

Whereas the other European neutrals can trace and date their neutralist genealogy almost to the day, this does not hold for Sweden. Thus, although the Swedish tradition of neutrality in foreign affairs has roots running back all the way to the demise of the Napoleonic period, it did not achieve its distinct contours until the end of the nineteenth century, when neutrality increasingly came to be regarded as the only feasible way for small states to resist being pulled willy-nilly into the vortex of the *fin-de-siècle* machinations of the major European powers.[10] As a foreign-policy option this choice was subsequently legitimized by the various Hague Conventions on the laws of warfare on land and sea, and the wartime rights and obligations of neutrals, signed in 1907.

The first real test of this policy came during the First World War. As it turned out, Sweden's neutralist stand was on the whole respected by both sides for the duration of the war, despite British pressure—contrary to the Hague Conventions and the practice of international law—aimed at curtailing trade between Sweden and Germany. The major effect of this British policy was to pave the way for a liberal–Social-Democratic coalition and, in general, a more pragmatic and Anglo-Saxon-oriented (rather than pro-Continental) Swedish attitude.[11]

Although some of the major premises on which the League of Nations was based were clearly incompatible with Swedish neutrality, Sweden nevertheless became a member in 1920. In justification of this decision, the government argued that all small states would sooner or later find it in their interest to join the organization, and that despite the obligation to participate in all economic—and even some military—sanctions imposed by the League (a requirement constituting a clear break with neutralist traditions), the advantages of membership in an organization dedicated to peace and disarmament out-weighed these and other drawbacks. Indeed, Hjalmar Branting, the founding father and first postwar leader of the Social Democratic Party, even suggested that Sweden should in the name of international peace and solidarity take the decisive step of *abstaining* from 'our time-honoured policy of neutrality, which has consisted of letting the world outside take its course while seeing to it that we are not involved in its conflicts, in so far as this is possible and allowed by the conflicting parties'.[12] This acceptance of the new order went so far that for an extensive period of time Sweden—in contrast to, for example, Switzerland—even stopped referring to her foreign policy in 'neutralist' terms. During this period the country also started to play an important role as a moderating influence and active proponent of disarmament within the League, as well as increasingly taking sides on international issues—the latter a stance differing fundamentally from the inward-looking neutrality of the pre-1914 years, at the same time as it augered the country's future activist bent and self-imposed role as a willing go-between in international conflicts.

However, after the disintegration of the League following the failure of economic sanctions against Italy in 1935–6—a League decision strongly supported by Sweden—the country once again reverted back to a more traditional neutralist foreign policy, albeit one that was less undifferentiated in its thrust than the neutrality relinquished upon entering the League in 1920. Rickard Sandler, foreign minister during most of the 1930s, at this time thus spoke of a 'Nordic' neutralism running parallel with Swedish neutrality. However, this regional dimension was not formalized by a defence pact or other type of joint security arrangement.[13]

Sweden was the only Scandinavian country to succeed in remaining neutral throughout the Second World War. This time around the pressures on her neutrality were, however, exceedingly more palpable, persistent and indelible. In fact, she was forced to compromise it on various occasions, foremost and most infamously as a consequence of a series of German demands following the occupation of Denmark and Norway, but later also in response to Allied pressure. Although the Swedish government was then and on later occasions variously pilloried for these compromises, it could claim not only that such a response was inescapable, given Sweden's precarious situation throughout the war, but also that the reason why Sweden *could* remain neutral lay not 'in the pursuit of strong-principled behaviour but precisely in the ability to adapt to circumstances'.[14] Thus, although the wartime government (consisting of all the major political parties) broke several of the articles of the Hague Convention, its position was that the limits of this 'adaptability' were always strictly determined in each instance by the exingencies of war, and hence that Sweden's prime consideration was never to accede to pressure except when this was absolutely necessary as a last resort for maintaining her territorial integrity and national sovereignty. By way of this argument Sweden also exerted herself to sustain throughout the war the confidence of both warring sides, however much the one would criticize her policies vis-à-vis the other.

The reasons why Sweden was able to stay out of the war are still being debated by Swedish historians. As Agrell has noted, no concensus obtains on the question whether neutrality prevailed due to the success of Swedish policy or as a consequence of a lack of incentives and/or resources on the part of Germany during the critical period between 1940 and 1943. Whatever the case may be, the following conclusions were drawn by Swedish foreign policy decision-makers:

> The main lessons of the war years . . . appeared to be that neutrality worked as a strategy for staying out of the war, that the balance of power between the combatants was of utmost importance and that the level of Swedish defence determined the extent of the room for manoeuvre for the government. The experience during the war thus created or confirmed major principles of Swedish security and these principles became the intellectual core of the postwar doctrine of non-alignment and a strong defence.[15]

These lessons, based on both the principle of foreign-policy feasibility (the 'politics of the acceptable', in Agrell's words) and on an extensive military build-up during the latter part of the war, have subsequently been forged into a central maxim of Swedish foreign policy: the utter necessity of combining a substantial element of pragmatic *flexibility* with the need, at all times, of maintaining—by means of principled and consistent action—the *credibility* of the policy of neutrality.[16]

'The Second World War', Krister Wahlbäck has written, 'left a much deeper imprint on Swedish experiences of neutrality than had the First World War. . . . The Swedish government had to pursue intensive negotiations with the warfaring parties to guarantee Sweden's survival as an independent nation and to retain the main contours of her neutrality'.[17] Consequently, having escaped the depredations of occupation by the bare skin of her teeth, it was not difficult for a country as isolated as Sweden to decide, once the war was drawing to a close, to participate actively in promoting the creation of a new international order and world organization dedicated to the principles of collective security: the United Nations. Indeed, this enthusiasm was so pronounced that Östen Undén, who would become Sweden's most important postwar foreign minister, as well as the architect of her present-day neutrality, declared in 1945 that the country was even willing to compromise—once again—some of the principles of neutrality if so required by the statutes of the United Nations.[18] However, since all the major powers had a veto in the Security Council, the chances of being forced to take sides in a conflict between the two blocs were thought by the government to be exceedingly slight. In addition, the country was now willing to forgo some of its previous neutralist policies if a defence pact, free of all larger alliance entanglements, could be established between Sweden, Norway and Denmark.

Very soon, however, the United Nations became a prime Cold War battleground, in which the use of the veto in the National Security Council came to symbolize precisely the development which it was the *raison d'être* of this organization—and the veto—to prevent: the division of the world into two blocs headed by two new superpowers. This also effectively undermined the establishment of a Scandinavian defence union, not only because Norway in particular now felt a stronger need to link her security in some way to a larger Atlantic system but also due to American pressure on all three Scandinavian countries to forgo a policy of neutrality in favour of a joint transatlantic defence of Western Europe. As to neutrality itself, this was certainly not a period when the United States—or, for that matter, the Soviet Union—gave much for the worth of what was regarded as both a morally questionable and politically infeasible policy within the context of the polarized struggle between Western and Communist values.[19] As a consequence of these and other factors, Norway and Denmark joined NATO in 1949 (although on the basis of so-called 'minimum conditions'), while Sweden decided once again to go it alone and to retain a policy whose official definition has remained constant since then, namely 'non-participation in alliances in peacetime aiming at neutrality in the event of war'.[20] Militarily, this policy was buttressed with a strategy of deterrence based on a large and costly defence system, which, it was hoped, would deny Swedish soil to a potential enemy except at what the latter would consider to be a prohibitive cost.[21]

Sverker Åström, veteran senior diplomat and authoritative arbiter of Swedish foreign policy, has latterly justified the choice of this doctrine in terms of the following Cold War and geo-strategic considerations:

> Sweden is not Chad or Uruguay, nor China or USA. Sweden is a medium-sized democratic country in northern Europe, with strong economic ties with the Western world, at the crossroads between major strategic interests, and in the immediate proximity of areas of vital demographic, economic and military importance to one of the superpowers. Being so situated, it stands to reason that Sweden should have two aims—the one, to avoid coming under the influence of the nearby superpower, and the other, to avoid becoming the menacing outpost of the other superpower. A policy of neutrality is the obvious answer.[22]

At least two additional reasons can also be adduced for this choice. The first is that in the view of most Swedes—ordinary citizens and decision-makers alike—the historical vicissitudes of the country since the early nineteenth century have time and again proven the preeminent worth of this policy. Or as Undén put it in an important speech in 1949:

The fact that Sweden has succeeded in keeping peace for 135 years has undoubtedly had a strong psychological effect on the Swedish people's view of her national security. During this long period Europe has been shaken by tremendous wars without implicating our country. Whatever the explanation for our success in evading war, it will not be easy to convince our people that their security now demands that we reject neutrality for being a wrong and outdated policy.[23]

That Sweden was willing in 1948–9 to exchange her own neutrality for a Scandinavian defence pact does not substantially affect this point, since what was envisaged was an agreement based on the continuation of neutrality, but now in terms of a larger and more powerful regional rather than national unit. As such it could be viewed as a logical extension of the *de facto* situation prevailing in Scandinavia prior to the outbreak of the Second World War. Second, it has often also been suggested—though mainly *ex post facto*—that given the NATO membership of Norway and Denmark, and Finland's Treaty of Friendship, Co-operation and Mutual Assistance with the Soviet Union, signed in 1948, Sweden was virtually obliged to remain non-aligned in order to create a necessary 'Nordic balance'—that is, strategic stability on the basis of a buffer state *cum* regional balance point—in postwar Northern Europe.[24]

Whatever the relative importance of these considerations, they have led all Swedish governments since the war to accept—indeed, to insist upon—the following two fundamental limitations on the country's freedom of action: (1) that '*no commitments* . . . be made in peacetime that prevent us from fulfilling the obligations of a neutral power under international law in a war between other states'; and (2) that 'in peacetime we must pursue a policy that inspires and *sustains the confidence* of the rest of the world in our determination and ability to remain a neutral and independent state in wartime'.[25] Although not explicitly stated, closer reflection shows that both of these tenets are based on the logic of precommitment previously adumbrated, although in different ways and with different consequences for Swedish foreign policy. The thrust of this claim, as well as the point of distinguishing between the two principles of action posited above, can perhaps best be appreciated if we think of them in terms of, respectively, the concepts of *sovereignty* and *autonomy*. I will argue below that Swedish neutrality is intimately linked to both of these hallowed notions of international law and politics; and that whereas the concept of sovereignty is in most instances unproblematic in this regard, this is not equally the case with respect to autonomy.

The above thumbnail sketch of the history of Swedish neutrality has also tried to indicate that this complicated nature of the doctrine is a relatively recent phenomenon, at least in so far as the pursuit of autonomy was not a prime foreign-policy consideration prior to the Second World War. It is also for this reason that it must be pointed out once more that Swedish neutrality has never been a cut-and-dried and hence unchangeable instrument of

Swedish foreign affairs. This documented capacity for adapting itself to changing circumstances may indeed constitute, if anything, a major explanation for its extraordinary durability.

Sovereignty, Autonomy and Neutrality

To clarify the distinction between sovereignty and autonomy is, as William Wallace has recently pointed out, 'not a straightforward task: economists, lawyers, politicians and political scientists all have their own preferred definitions'.[26] However, he then cites a set of definitions—appropriated from Fred Hirsch—which he finds useful in this connection: '*Formal sovereignty* concerns legal rights. *Effective sovereignty* concerns the practical power to exercise those rights. *Autonomy* concerns the results achieved by their exercise'.[27] To this Wallace then adds the following comments:

> Sovereignty thus looks in principle to be a clear and measurable concept. Either a government has it or it does not; when it yields up a portion of its sovereignty to some international authority, it yields precisely that and no more. *Autonomy*, by contrast, is a much more relative concept, subject to the shifting impact of domestic and international developments; the limits to national autonomy may be exposed only under conditions of crisis, as through the run on sterling that followed the Suez intervention in 1956.[28]

A similar distinction is to be found in Richard N. Cooper's conception of sovereignty as 'the formal ability of a nation to act on its own rather than under the instruction of another nation', in contrast to national autonomy, which 'is the ability of a nation to attain its objectives through unilateral action'.[29] The relevance of these distinctions to an understanding of Swedish neutrality is, in my view, the following.

When Åström writes that 'no commitments . . . be made in peacetime', he has in mind commitments which have the effect of, in Wallace's terms, 'yielding up a portion of [Sweden's] sovereignty to some international authority'. In other words, he is here referring to that aspect of Swedish neutrality which involves a strict precommitment *not* to be swayed by *any* Siren's song in the form of either international alliances, agreements or guarantees, however appealing they may sound to the ears of a small country left, admittedly by choice, to her own devices on the high seas of world politics. This is not only a very precise and measurable policy but one which also differs from those of other European neutrals, all of whom are in some sense or other signatories to international accords specifying and/or guaranteeing their neutralist status. Åström has given the following reason why Sweden has not followed suit:

> Sweden has rejected the idea of incorporating neutrality in any international agreement. In our view, guarantees furnished by the great powers would create some

measure of dependence on those states. They might claim the right to keep an eye on Sweden's foreign policy and raise objectives should they consider that it conflicts with the terms of the international guarantees.[30]

However, although Swedish neutrality is in this sense not regulated in an international agreement of any kind, this self-imposed structural constraint at the same time contains the explicit implication that this policy—precisely because of this strict interpretation of non-alignment—will lend special force to Sweden's expectation of being treated as a *bona fide* neutral power under international law in the event of warfare beyond her borders.[31]

This dependence on international law and agreements only during conditions of war may at first seem paradoxical; it becomes somewhat less so in view of the absence of any international rules of jurisprudence governing the proper behaviour of neutral states during peacetime. Indeed, since 'neutrality' is pertinent only to conditions of warfare, Swedish governments have always argued that there is nothing in her foreign-policy doctrine which obligates Sweden to be 'neutral' in an ideological, cultural, economic or other non-military sense in her peacetime dealings with foreign states or in her domestic political behaviour. The sovereignty which is at issue here, and which it is the aim of this policy to secure and to protect, guarantees her this very right.[32]

However, when Åström goes on to specify the second limitation on Swedish freedom of action, which refers to the creation of international confidence in Sweden's ability to remain fully neutral in wartime, he is no longer speaking of state sovereignty but of *policy autonomy*: the ability of a state to achieve the objectives which it has set for itself within an issue area. Since this is a highly relative and dynamic notion, it is not measurable in the 'either/or' sense that aspects of the sovereignty of a state can be determined. This follows from the view (*vide* Hirsch's definition above) that sovereignty concerns the power *to* exercise rights, while autonomy pertains to the effects *of* this exercise. Thus, while both concepts refer to a *power relationship*, the former implicates power in the formal sense of 'rights' (which a state has or has not been able to appropriate), while the latter does this in terms of 'causality'—the 'production of intended effects', to cite Bertrand Russell's classic and felicitous definition of power.[33]

Before analysing what it means more specifically to regard Swedish neutrality in terms of the concept of power, let me illustrate both the appositeness and the importance of the above distinction by way of the following example. In a book on Swedish 'civil' national security, Agrell has described Swedish neutrality as follows:

> The policy of neutrality is built on the assumption that Sweden is in all essential respects *able to determine* her position in the international system. A persistent trait of this policy has been precisely this emphasis on Sweden's *independent determination* of her policy, on her *rejection* of all international guarantees, and on her *exclusive responsibility* for the defence of the country.[34]

The emphases in the above passage are mine, and have been added for the purpose of showing that these italicized predications can, broadly speaking, refer either to (1) *aspects of state sovereignty* as discussed above, in the sense that Sweden maintains that she has sovereignly been able to act thus with respect to the substantive issue areas indicated in the passage, or to (2) *aspects of policy autonomy*, that is to claims regarding the actual exercise of power in respect to these issues. In short, the phrase 'able to determine' can refer either to the appropriation of sovereign rights or to the effective exercise of such rights. And while the former *qua* doctrine does not implicate the latter *qua* policy behaviour (which rests on empirical, not logical, determination), it follows from the definitions posited above that the effective exercise of rights presupposes the right to exercise them.

Two immediate issues are pinpointed by this interpretation of the above passage. First of all, our example illustrates the essentially dual nature of neutrality, and hence the need, at all times, to distinguish between these two fundamental dimensions of Swedish foreign policy. Second, it makes a fundamental difference which meaning or dimension is implied when discussing and evaluating this policy. For while, empirically, the first interpretation above is politically uncontroversial and analytically unproblematic, a similar response to the second would in most instances constitute a prime case of begging the question. It is not at all certain that Sweden is today in fact able to exercise effectively the rights, or doctrine, which she has appropriated by declaring herself a non-aligned state in international relations. This is, after all, precisely the issue raised by the so-called submarine crisis discussed above; and an appeal to manifestations of Sweden's 'sovereign' (as distinguished from 'autonomous') power will certainly not resolve it. As Kenneth N. Waltz has reminded us, to be 'sovereign and to be dependent are not contradictory conditions'.[35] The same cannot be said about autonomy.

Without making the above distinction, Agrell implicitly recognizes this conclusion when he adds the following significant passage to the description of Sweden's neutrality quoted above:

> The actual implications of the policy of neutrality have, however, changed, but not according to a conscious political strategy. The development of military technology has meant that Sweden—in order to be able to maintain her military objectives—has become increasingly dependent on other countries. The same applies to large sectors of the economy, simply because of the internationalization of the Swedish economic system. In the beginning of the 1980s a series of incidents have pointed to this conflict between *conceptions* of the means available to Sweden for pursuing a policy of neutrality according to her own choice, and the actual means available for so doing. It is symptomatic that so few have been willing to pose the decisive question, viz., if the preconditions for the established policy of neutrality have changed.[36]

He here at the same time intimates a fundamental difference between how *changes* in these two types of power relationships in international relations are effectuated. Thus, on the one hand, changes in the sovereignty of a state are a consequence of dramatic events, such as wars, revolutions or referenda, when—willingly or unwillingly—sovereignty, or an aspect of sovereignty, is *transferred* to another sovereign body. Examples of each abound: looking at Sweden's closest neighbours, for example, Norway lost the sovereignty she had gained in 1905 as a result of German occupation in 1940; the Bolshevik Revolution transferred sovereignty to the newly independent Finland in 1918; and Danish membership in the European Community was a consequence of a national vote on this issue in 1972. These examples also indicate why this concept is so eminently measurable, since in each instance it is essentially a question of either yes or no, at this or that point in time, referring either to sovereignty as such or to selected aspects of it.

In contrast, changes in the autonomy of a state is a matter of incremental, often unanticipated and usually unwanted change: the kind of change which, in Karl Popper's view, constitutes the only social or political subject-matter really worthy of scientific inquiry, precisely because it is not based on clear, measurable intentions (which call for 'mere' description, not explanation) but on the unintended consequences of a host of actions, often in complex and largely impenetrable societal configurations.[37] This also explains why, although potentially a constant occurrence, change in policy autonomy is such a difficult phenomenon to measure, both generally and in particular instances. For example, how are we to determine if a policy such as neutrality actually has an intended or purported effect in international relations at a given point in time?

As I have already intimated above, a discussion of this issue must proceed in terms of the concept of power—the genus of which autonomy is a species. But what does it mean to say that neutrality is a power relationship and that the neutrality of a state can be eroded as a consequence of losses in its autonomy? In my view, this query raises the most difficult problem in the analysis of Swedish foreign policy today: how to respond to Agrell's 'decisive question' above, namely whether the preconditions for Sweden's neutrality have or have not changed since the wartime situation which, it is generally agreed, provided the present Swedish policy of neutrality with its essential ingredients. Too often the answer to this query has been effectively fudged by the failure to pose it with specific reference to what is, after all, the nexus of all foreign policy: the question of power itself. Once this is understood, I will argue below, it will also become clear why issues pertaining to the credibility of a policy such as neutrality are essentially of a structural and substantive rather than cognitive nature.

Power, Influence and Neutrality

Although the notion of power is both ubiquitous in the literature and frequently contested as an essential tool for understanding interstate relations, this is not the proper place to enter into an extensive discussion of its analytic merits in the study of foreign policy. Instead, I shall forthwith accept Felix Oppenheim's assumption that 'this notion is indispensable for political analysis, not only in its own right but also . . . because it belongs to a network of other key notions such as influence, deterrence, coercion, punishment, authority, legitimacy, and also unfreedom and freedom'.[38] My aim here is to argue that the policy of Swedish neutrality involves a power relationship of a particular kind; and that very specific implications are contained in this claim—implications which can be clarified by means of a *conceptual reconstruction* of this policy in terms of the concept of power.

The validity of the claim that neutrality is not only a doctrine but also, and more importantly, constitutes a power relationship, can be gauged by means of semi-official declarations of the following kind, culled from Åström's discussion: that it is the intended function of this doctrine 'to strengthen the safeguards of our national independence and our democratic society, to preserve and, if possible, improve the chance of saving Sweden from war'; that this policy expresses 'our determination and ability to remain a neutral and independent state in wartime; or, more specifically, that 'Sweden should have two aims—the one, to avoid coming under the influence of the nearby superpower, and the other, to avoid becoming the menacing outpost of the other superpower'.[39] In my view, these declarations clearly have the typical qualities of 'power' as this notion has been conceptualized and applied in modern social science since Harold Lasswell and Abraham Kaplan first published their seminal study on *Power and Society* in 1950.[40] Let me attempt to clarify this assertion as follows.

The trademark of this modern conception is the notion of causality, in contrast to the traditional view of power—stretching from Niccolò Machiavelli to Hans Morgenthau—postulated as an innate or acquired property of the power holder. David A. Baldwin has epitomized this contemporary conception in his assertion that when discussing power 'it is essential to specify or at least imply who is influencing whom with respect to what; in short, both *scope* and *domain* must be specified or implied'.[41] This echoes Robert A. Dahl's classic and equally unambiguous position in this regard:

> Any statement about influence that does not clearly indicate the domain and scope it refers to verges on being meaningless. When one hears that A is highly influential, the proper question is: Influential over what actors with respect to what matters? The failure to insist on this simple question often leads political observers astray.[42]

Following Harold and Margaret Sprout, Baldwin has also referred to this necessary specification of the causal parameters of power in terms of the broader notion of a 'policy-contingency framework', signifying the essentially contextual nature of all social manifestations of power.[43] In the passages on neutrality quoted in the previous paragraph, these characteristics of the concept of power are easily specified in terms of the following two predications: Sweden trying to influence the superpowers to respect her neutrality in the event of war. The first of these refers to domain—the target of attempts to influence. The second refers to scope—the objectives of the attempt in question. It is with reference to these elementary conceptual attributes of power that we can speak of Swedish neutrality as a *policy* aimed at implementing, sustaining and, hence, *ipso facto* constituting a particular form of *power relationship* defined in terms of a specific contextual structure; and why, whenever we wish to investigate the manifest function of this policy in Swedish foreign affairs, it must be couched in terms which are directly related to this policy-contingency framework of Swedish international relations.

However, given the complex nature of the notion of power, the concept is in need of at least some further analysis before it can be utilized for our specific purposes. The immediate question here is the following: what type of power relationship is implicated when speaking of Swedish neutrality? Following the main outline of Oppenheim's discussion of the inclusive concept of social power, I shall here argue that pursuing a policy of neutrality is an instance of *exercising influence* in international relations. This broad claim must, however, be qualified in terms of at least the following considerations.

First of all, when speaking about influence we are referring to a specific type of power relationship, not to the concept of power itself *in toto*. Thus, as Oppenheim has argued in some detail, influence should be clearly distinguished from both the power to *coerce* and the power to *punish*, both of which not only constitute important types, or subcategories, of power but which also figure prominently in the international-relations literature. Sweden's neutrality differs from much of international politics precisely in its rejection of both coercive and punitive policies in favour of a policy of exercising influence.[44] Second, the exercise of influence must be understood as a subcategory of the notion of *having* influence, since one can have influence without exercising it, but not vice versa.[45] For reasons which will become evident below, I have preferred to conceptualize neutrality in terms of the less inclusive of these two notions of influence, that is as a way of exercising rather than possessing influence.

Third, although power, and hence influence, can have both a positive and a negative denotation, it is the latter which is most germane in the present context. Swedish neutrality is, if anything, a policy aimed at *preventing* the occurrence of certain types of actions during conditions of war. In other words, when discussing this policy our focus is on policy actions involving the non-performance of specific types of action as a result of the exercise of a power relationship. Oppenheim has formally defined this form of interaction as follows:

Definition: P influences R not to do x if and only if P performs some action y that causes R to choose not to do x.[46]

Fourth, it should also be made clear that influence as discussed here refers to influence between single actors, not to situations of collective power, either of the kind where 'an individual can be compelled to act in a certain way not by one or more actors, but only jointly', or where action is directed toward a joint power constellation, such as an alliance, a balance-of-power system or a nuclear deterrence structure.[47]

Fifth, the relatively low fungibility of political power—and hence influence—also needs to be emphasized here, especially when compared to economic power resources. As noted by Baldwin:

> Any particular economic power resource can usually be converted into another kind of economic power resource. Money, as a highly liquid medium of exchange that also serves as a standard of value, facilitates such resource conversions. In the political realm, however, there is no close counterpart to money; therefore, it is much more difficult to convert one kind of power resource into another. This lack of fungibility of political power resources, together with the multidimensional nature of power relations, increases the probability that an actor may control large amounts of potential power with respect to some scopes but relatively small amounts of potential power with respect to other scopes. Thus, a nation may be powerful with respect to deterring nuclear attack but weak with respect to getting one of its citizens elected Secretary-General of the United Nations.[48]

This is precisely why specification of scope and domain is so essential when discussing power and influence in international relations, and why policy analysis must be contextual, taking account of the power actually manifested in each concrete political situation, as Lasswell and Kaplan noted many years ago.[49]

Finally, although influence refers to unilateral power relations, this does not mean that one cannot extend the analysis to include either reciprocal but asymmetrical relations between two actors (when one actor has influence over another in one respect, and the latter has influence on the former in some other respect), or relationships involving three (or more) actors at the same time. Indeed, neutrality as discussed here is precisely such a policy, since it by definition involves at least three actors, two of whom are in conflict with each other, while the third's policy it is to be aligned to neither.

These last two points can be illustrated by means of two simple figures (Fig. 4.1), both intended to represent influence relationships between three actors in an international system, two of which are involved in a conflictual relationship with each other (the two superpowers SP1 and SP2), while the third (the neutral actor N) is neither aligned to, nor in conflict with, the other two.

Figure 4.1(a) represents an example of a neutralist state attempting to affect the external conditions of her neutrality—and international relations more broadly—by means of influencing the *interaction patterns* of the major international actors, and hence structural aspects of the international system itself. Many of the activities of the neutral states during this century have been of

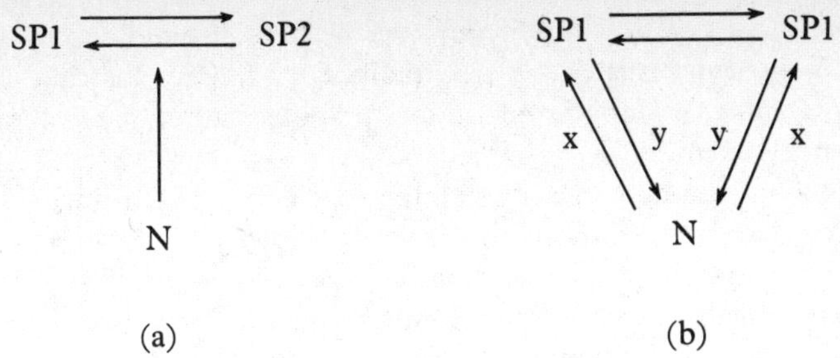

Figure 4.1 Two types of neutrality

this kind, essentially involving attempts to establish international agreements, rules, structures or 'regimes' favouring, in some manner or other, the sovereignty of states of the neutralist persuasion. Using a term from Elster's discussion of rationality, this can also be called a parametric policy of neutrality, a view equating the pursuit of security with the pursuit of peace, that is with a policy aimed at changing the given structural and operational principles (or parameters) of the larger world system in order to lessen international tension.[50] In my view this is not, however, the type of interaction which has characterized the logic of Swedish neutrality as practised since the Second World War.

The latter is, instead, illustrated by Figure 4.1(b) posited above. Its function is to indicate that not only must we in this connection think in terms of a power relationship between Sweden and *both* of the superpowers *at the same time*, but that *two* types of interaction patterns are involved in each such reciprocal, asymmetrical relationship. The purpose of x in each instance is to influence both SP1 and SP2 to do y, that is to refrain from violating Sweden's neutrality in the event of war.

Thus, although couched in terms of Oppenheim's explication of the concept of social power, this figure at the same time represents the most essential elements of the Swedish policy of neutrality itself as discussed above. In contrast to a parametric policy, this is thus—using Elster's terminology once more—a quintessentially strategic approach to international relations.[51] In other words, rather than aiming to reduce international tension, this approach focuses exclusively on an actor's pursuit of autonomy.[52] This figure also shows in what sense one can claim that Swedish neutrality is based on a 'realist' rather than 'idealist' conception of the requirements of autonomy in the international system, namely that neutrality must perforce—if it is to move beyond aspects of state sovereignty to aspects of foreign-policy autonomy—be pursued on the basis of one-to-one influence relationships with both of the leading world powers. In other words, the doctrine of Swedish neutrality has *per se* nothing to do with 'improving' the international system or making it more compatible with the interests of neutral (or other) states. Rather, despite the ameliorative aura which is often claimed to emanate from this policy (or, perhaps, from some of its more unctuous representatives), it is essentially a hardnosed doctrine aimed at maximizing and sustaining Sweden's foreign-policy autonomy in the present international system.[53]

The contents and outcome of the intense polemical battle during the first postwar decade between Foreign Minister Undén and Herbert Tingsten, famous editor-in-chief of *Dagens Nyheter*, bear witness of, if anything, this fundamental 'realism' of Swedish foreign policy as practised by its chief architect. Very briefly, Undén's policy was to make a clear distinction between questions pertaining to, respectively, Swedish national security, international disarmament and the pursuit of human rights and democratic principles in the international arena. Neutrality he regarded as a policy solely of national security, and as such it is to be pursued independently of all other international and normative considerations. Tingsten, on the other hand, adamantly refused to view the above as three watertight policy areas. Rather, in a thoroughly Wilsonian fashion he argued that questions pertaining to the first two should always be made subservient to a fundamental Swedish commitment to human rights and democratic values in her international interactions. This would, of course, have made impossible a continued commitment to Swedish neutrality, an implication which Tingsten was fully aware of.[54]

Although I have not explicitly argued in such terms above, maintaining a structure of interaction as described here calls for a dynamic feedback system of the kind illustrated in Figure 4.2. In short, the Swedish exercise of influence—as manifested in the superpowers' respect for her neutrality—requires having influence, which is based on the possession of autonomy; and autonomy, in turn, is a function of the successful exercise of influence. In this logic the concept of influence is thus central for a very simple reason: without the ability to exercise influence, autonomy is—by definition—an empirically hollow concept. Although this may at first seem to be a point merely of formal-logical interest, it is in fact crucial to answering the 'decisive question' posed by Agrell above.[55] Let me explain why by referring, once again, to Oppenheim's discussion of social power.

Given the formal definition of influence posited above, it follows, according to Oppenheim, that 'P exercises influence over R's conduct *to the extent* that R's decision is the result of some influence action of P'. Hence, if 'R decides to abstain from x independently of P (on his own or as a result of another actor's influence), P cannot have influenced R's conduct—"cannot" in the logical sense, given the proposed definition'.[56] In other words, in order

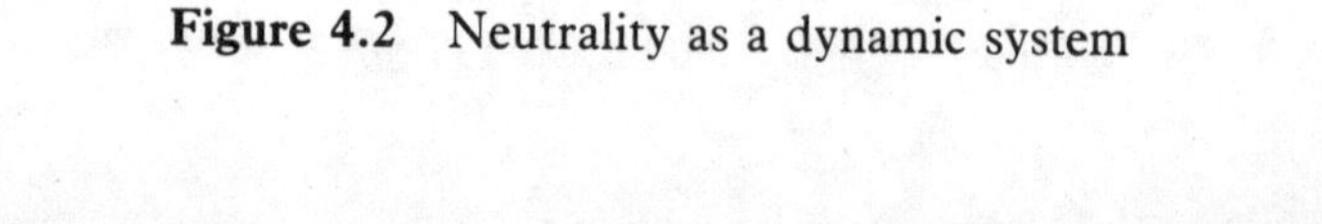

Figure 4.2 Neutrality as a dynamic system

to be able to examine *if* a neutral country such as Sweden is in fact able to act autonomously in her foreign undertakings vis-à-vis the superpowers, it does not suffice to note that the latter have hitherto agreed, in word or in deed, to respect Swedish neutrality in the event of war. For it is not only logically but also empirically entirely feasible to hypothesize that the superpowers may have agreed, or may in future agree, to do so *for other reasons* than the successful exercise of influence on the part of Sweden. Clearly, this is a very disturbing thought, not only due to the suggestion of duplicity and tergiversation which is implied by such an hypothesis, but also because such putative reasons may quite conceivably be connected with factors over which a country such as Sweden would, even if she were aware of them, have little or no influence.

This type of uncertainty—and the problems of mutual distrust underlying and sustaining it—is, of course, an endemic feature of all international relations, and has as such become a factor which most states have of necessity learnt to live with over the years (and centuries). Nevertheless, for a Lone Rider such as Sweden it is particularly important to counteract its potential effects as much as possible, especially if these are liable to be compounded by a faulty or inadequate understanding of the analytic presuppositions and logical implications of a policy such as contemporary Swedish neutrality. Hence we need to discuss further the nature of the dynamic notion of interaction illustrated above, and in particular the question as to *how*, in fact, a neutral state of Sweden's size and geo-strategic location is at all able to exercise influence on the two superpowers. In other words, what means are available for maximizing whatever foreign-policy autonomy is available to a small state such as Sweden?

Here, once again, Oppenheim's discussion is of great value, both heuristically and as an explicatory framework for further examining the conceptual ramifications of the notion of influence. He thus notes that in general 'P's influence action y usually consists of some verbal communication by which he *persuades* R to choose a certain alternative or *dissuades* him from adopting some course of action'.[57] As explained by Hiller Steiner, this implication of 'influence' is by no means fortuitous:

> Offers and threats are interventions, by others, in individuals' practical deliberations. They are intended by their authors to influence how a recipient individual behaves, by altering the extent to which he actually desires to do a particular action of a kind which he is able to do.[58]

Furthermore, the actual practice of Swedish neutrality clearly substantiates the applicability of this analysis, based as it is on a very simple dialectical logic, combining both persuasive and dissuasive elements: that Sweden will allow neither of the superpowers to threaten the other by means of Swedish military, territorial or other help.[59] In other words, to be able to *persuade* both of the superpowers of their mutual advantage in accepting this state of affairs,

Sweden is forced *pari passu* to *dissuade* both from attempting to 'use' her against the other.

However, it must immediately be emphasized that although neutrality is based on the use of both positive and negative sanctions (here I revert to Baldwin's terminology), these two approaches are not of equal dignity as instruments of persuasion. Rather, I shall here argue that it is the former rather than the latter type of influence which lies at the heart of Swedish neutrality. As Baldwin has argued, 'positive sanctions are significant resources by which international actors affect the behaviour, beliefs, attitudes or policies of other actors'; and this, as I have maintained above, applies with special force to Swedish foreign policy.[60] However, although it is here claimed that neutrality is essentially a relationship based on the efficacy of positive rather than negative suasion, it must immediately be added that actions of the latter kind nevertheless constitute an essential component of the former. In short, rational dissuasion (or deterrence) is not as such an instrument *of* Swedish neutrality; its use is, however, a prerequisite as a lever *for* the effective exercise of this policy. The logic underlying this claim can be clarified as follows.

The fulcrum around which Sweden's relationships with the two superpowers revolves is, as pointed out above, the promise-*cum*-threat that she will let neither of the superpowers use Sweden against the other. This double commitment constitutes a *strategic good* from the point of view of both superpowers, and has as such formed the basis for Sweden's relations with them since the Second World War. However, the efficacy of this dual relationship is a function not only of the mutual benefits provided by this exchange but also of the persuasiveness of this ongoing commitment: Sweden's actual ability to rebuff any and all attempts from the side of either superpower to use her against the other. Hence the necessity here—both logically and empirically—of negative sanctions in buttressing a relationship based on positive sanctions in the provision of this particular strategic good; and why, although neutrality as an instance of international behaviour is constituted by the *exercise* of influence in the form of rational persuasion, it at the same time presupposes the constant availability of *sources* of negative influence, that is the *ability to make credible* the threat that all attempts to undermine her sovereignty will be repelled.

In this connection it is important to note, however, that since it is the threat of a negative sanction directed at the *one* superpower which undergirds the effect of positive sanctions directed at the *other* superpower, and vice versa, the logic sketched here does not exemplify a stick-and-carrot approach to international relations. Such an approach on the part of a neutral state vis-à-vis the superpowers would, of course, be exceedingly foolish. Rather, Swedish neutrality is a complex instance of basing the exercise of influence on the use of positive sanctions, while implicating negative sanctions solely for purposes of undergirding the feasibility of such relationships in international affairs. Hence, although it involves reference to the potential availability of instruments of dissuasion, the implications of the logic discussed here are essentially non-conflictual.

Although I shall not discuss this issue in any detail here, the importance of *diplomacy* as the central instrument of Swedish foreign policy should also be emphasized at this juncture of the present analysis. In his book on *Economic Statecraft*, Baldwin makes a distinction between four fundamental techniques of statecraft in international relations: propaganda, diplomacy, economic statecraft and military statecraft.[61] On the reasonable assumption that these are mutually exclusive and jointly exhaustive categories, the thrust of the discussion so far is to indicate, if anything, the essentially diplomatic nature of Sweden's undertakings vis-à-vis the two superpowers.[62] This follows not only from the nature of Swedish foreign policy as discussed above but also—by a process of elimination—from the fact that none of the other three types of techniques is pertinent when discussing the concept of Swedish neutrality (as distinguished from foreign policy more broadly conceived). This also explains Sweden's consistent and genuine dedication to multilateral international organizations, since these—such as the United Nations—are prime forums for the type of statecraft on which Swedish foreign policy is premised.[63]

One further aspect of the use of persuasion and dissuasion as influence techniques should be noted here. Oppenheim has argued that these two types of interaction patterns may in principle be based on the provision of either correct or misleading information, allowing the influencing actor a free choice between, on the one hand, rational forms of argumentation or, on the other, affective, manipulative and kindred methods of international discourse. However, a closer reflection on the nature of the particular kind of power relationships discussed here leads to the conclusion that a neutral actor such as Sweden is *a priori* obliged to exclude the use of all forms of subterfuge, deception or manipulation in her diplomatic dealings with either of the superpowers. Diplomacy as the influence technique *par excellence* of the neutral state is based on the feasibility of policies of *rational* persuasion and dissuasion; hence its successful pursuit—that is, the provision of the strategic good on which neutrality is premised—will depend on a continuous reinforcement of the principles of trustworthiness, openness and good faith underlying the effective exercise of persuasive rationality. The use of 'reasons of state' in the realist (and pejorative) sense of the term is, in other words, not compatible with a neutralist policy of the Swedish kind. Instead, being based on bilateral power relationships with the superpowers, this policy must be premised on what, in K.J. Holsti's view, constitute the particular characteristics of all such relations, namely trust, reputation and role behaviour.[64]

Although this restriction may, once again, suggest a normative, idealist tenet in some sense or other, it is nothing of the sort. This constraint is simply a fundamental implication of the logic of rational persuasion on which—it is here claimed—Swedish neutrality is based. The last thing a country such as Sweden would want is to find herself in a Prisoners' Dilemma vis-à-vis either or both of the superpowers, the logic of which would propel all concerned towards non-cooperation and hence, in this particular case, to the devaluation of Swedish neutrality as a strategic good from the point of view of the two superpowers. To forestall any possibility of such an undesirable loss of influence, it is imperative not only that all forms of underhanded stratagems be avoided altogether, but also any *suggestion* even of 'tactical' behaviour on the part of the neutral state. Amartya Sen, in his analysis of the so-called Assurance Game, has shown that if full information is made available to all players in a situation of the kind discussed here, they will tend to co-operate in the confident expectation that everyone else will do the same, since no actor has anything to gain from acting differently.[65] The suggestion here is, in other words, that as long as no uncertainty or suspicion prevails with respect to Sweden's foreign policy intentions, the preferences of both superpowers will tend to reflect—everything else being equal—the Assurance Game rather than that of the Prisoners' Dilemma. The ambition of creating and sustaining this kind of collective rationality is, in any case, the rock onto which the logic of Swedish neutrality has been cemented for over four decades.

Vulnerability, Credibility and Neutrality

However, it will not suffice simply to play an honest hand; it has to contain good cards as well. The ability to exercise rational persuasion through diplomacy is based on the possession of power resources commensurate with the persuasive effects aimed at, that is the effects valued by the superpowers (in this case, in the form of a jointly valued strategic good). In other words, when Sweden maintains that she will allow neither of the superpowers to obtain a Swedish bridgehead against the other superpower, or to violate Swedish sovereignty in some other way which may be regarded as detrimental by either, this means that she must have sufficient defensive power resources to back up and give credence to such a promise-*cum*-threat. In other words, she must have the kind of *power base* which is a prerequisite for maintaining her policy of neutrality. What does this mean?

Baldwin has defined a power base as 'the causal condition that makes a successful influence attempt possible'.[66] The most important distinction underlying this conception is the difference, discussed above, between property and relational concepts. Contrary to what is often claimed in the literature, a power base is not a property of a single state, that is a factor—such as population, wealth or military prowess—which can be defined and measured without reference to other states. Baldwin thus notes the following:

> In contrast to such properties are terms like influence, capabilities, power bases, and power resources. These terms are relational in the sense that they say something about

an actual or potential relationship between two or more actors. It is impossible to describe actor A's capabilities without saying or implying something about the value system of actor B.[67]

Nevertheless, this if often done in the literature. One consequence of treating property and relational concepts alike is that statesmen 'are often described as "employing" or "using" their "capabilities" or "power bases" as if these were possessions of a single state'.[68] In view of what has been noted here, it follows that when speaking of Sweden's power bases we are here *not* referring *per se* to her wealth, military power or other similar national attributes. Rather, in all such instances we aim to invoke precisely those capabilities which have made—and have the potential of making—the exercise of influence on the superpowers effective. And these, as we have seen above, must be conceived of in the contextual-relational terms of the whole policy-contingency framework of Swedish neutrality delineated here.

In this context one further aspect of the power component of Swedish neutrality should be pointed out. Richard K. Ashley, in his spirited attack on the neorealist conception of power, has referred to the 'competence' model of social action, a notion with roots running back all the way to Thucydides's 'Melian Dialogues' but with implications which, in his view, are equally pertinent today.[69] The central thrust of this model is the idea that the power of an actor depends on, and is limited by, the conditions of its *recognition* within a larger community of actors. Ashley expands on this notion as follows:

To have power, an agent must first secure its recognition as an agent capable of having power, and, to do that, it must first demonstrate its competence in terms of the collective and coreflective structures (that is, the practical cognitive schemes and history of experience) by which the community confers meaning and organizes collective expectations. It is always by way of performance in reference to such collectively 'known' (but not necessarily intellectually accessible) generative schemes that actors gain recognition and are *empowered*. Thus, according to a competence model, building power always has a community-reflective performative aspect.[70]

Here, in my view, we have the linkage which is required in order to make credible the claim that the logic of neutrality explicated here has an empirical anchorage in contemporary international-political reality. That is to say, if the logic of Swedish neutrality works at all, it does so because Sweden has in fact been 'empowered' by the international community of states—and, in particular, by the two superpowers—to act as a specific type of actor, that is as the neutral type discussed here. By the same token, such power has obvious limits in the sense that such an actor can never with impunity exceed the limits of the recognition in question; nor is such recognition given once and for all. Ashley refers to this dynamic tension in terms of the 'dialectic of power and recognition'; and, as we shall see shortly, it is with reference to this dialectical process that the notion of credibility obtains its special meaning as an essential component of the logic of neutrality.

For purposes of the present discussion it is only at this point—and hence not in the explication of the logic of Swedish neutrality itself—that the issue of *size*, broadly conceived, first enters into the analysis. There is a good reason for this. In my view, the distinction between Sweden as a *neutral* state and Sweden as a *small* state is far too often fudged, ignored or downplayed in the literature.[71] Neutrality has as such nothing to do with smallness, nor smallness with neutrality. It is in order to avoid any temptation of conflating the two that this factor has been ignored until now. For the same reason we should also distinguish between the scope and domain of Swedish neutrality on the one hand and, on the other, issues pertaining to the *weight* and *cost* of the power base required to maintain this power relationship with the superpowers.[72]

Given both the low fungibility of power and its multiple and intertwined sources, it is obvious that no absolute measures are available for determining the values of these latter two variables. Indeed, one of the fundamental reasons for the inherently unstable dynamism—despite contrary appearances—of a policy such as Swedish neutrality is that its effectiveness over time is entirely dependent on the constant calibration and optimization of these factors in terms of both international changes and domestic political and economic considerations, that is in terms of the *whole* contextual framework of foreign policy. The problem constantly facing Swedish decision-makers in this regard is thus both classic, constant and formidable: how much and at what price? These questions become particularly acute if Swedish autonomy is being threatened not only because of superpower behaviour around her borders but as a result of her growing dependence on various external commodities, products, international regimes and other similar structural-economic factors of vital importance to her national well-being.

This brings us to the role of the notions of interdependence and vulnerability in the analysis of Swedish neutrality. What does it mean to claim, for example, that Swedish neutrality is being undermined as a consequence of the growing interdependence of the industrialized nations, or that this policy is in danger of being debilitated because of, say, energy vulnerability? I shall here argue that claims of this kind—whether valid, untrue or simply muddled—should *not* be approached by way of an analysis of Swedish foreign-policy behaviour as such, that is with reference to either official declarations, foreign-policy doctrines or, more broadly, the verbal give-and-take between Sweden and the superpowers. Nor, for the same reason, should they be gauged in terms of what is said, officially or otherwise, in domestic political debates. On the contrary, as the preceding discussion has tried to argue, it is here claimed that the only feasible way to proceed is to make a determination exclusively in terms of the *power base of neutrality*.

The reason for this focus on capabilities rather than on doctrinal discourse, official diplomatic declarations or political rhetoric in general is of a twofold character: an actor (such as Sweden) may inadvertently become increasingly vulnerable to environmental change without being, or being made, aware of it (here the notions of misperception and misinformation as discussed by Robert Jervis are highly pertinent); but an actor can also be induced to believe in such a development even though it has not occurred, for example, purely for manipulatory reasons (this, too, involves various forms of faulty cognitive processing).[73] In the first instance, vulnerability is a creeping but unrecognized fact; in the second, it is imputed and believed to be a fact without 'in fact' being one.[74] In either case, it is imperative—for the decision-maker and scholar alike—to counteract misperceptions of this kind by distinguishing between the sources and the actual exercise of power, that is between the *power base* and the *influence* for which the former is a causal condition. That is to say, whereas the specific nature of a power relationship in international affairs is always a function of the exercise of influence (or of some other form of power) at a given point in time, vulnerability refers to a structural-contextual condition underlying—and, from the scholar's point of view, analytically distinct from—the exercise of such influence.

In a previous chapter the concept of vulnerability was defined in terms of the opportunity costs of severing a given structural relationship between a national actor and the international environment. The higher the cost of such a disengagement, it was argued, the higher also the level of vulnerability is conceived to be. Given this criterion and measure of dependence, it also follows that the more vulnerable Sweden becomes to the one or the other superpower, or to actors and conditions controlled or significantly influenced by either, the weaker—given a *ceteris paribus* clause—becomes the power base of this policy. And clearly, the weaker the power base of Swedish neutrality, the less credible will her policy of neutrality become. It is in this sense that vulnerability is inversely related to the worth of Swedish neutrality as a strategic good in superpower relations. As intimated earlier, this conclusion has nothing to do with stated intentions, cognitive factors or other attitudinal variables: it hinges exclusively on changes in the power base itself, defined in the structural, objective and measurable terms of the policy-contingency framework which has been discussed here. Also, it is only by utilizing a conceptualization of this kind that we can respond properly to the following admonition, containing what is undoubtedly a sound precept but little in the way of practical advice:

> If we are to retain the autonomy required by the neutrality doctrine, we must learn to distinguish better between the real and the tactical criticism directed against our policy. It is only through respecting the former and dismissing the latter that the Swedish policy of neutrality can secure maximum credibility and prestige.[75]

This view corresponds well with Bo Johnson Theutenberg's recent dictum that the 'purpose of Sweden's peacetime policy is to build up and maintain the *credibility factor* with respect to Sweden's neutrality during war'.[76] Or rather, such an interpretation must, in my view, be the implication of his emphasis on the following two factors in this context: 'the *will* and *capacity* of the neutral state to uphold neutrality'.[77] Clearly, 'will' and 'capacity' are power base concepts *par excellence*.

This is a specification of the meaning and implications of credibility which differs considerably from the usual way that this factor is conceived by most Swedish analysts and decision-makers. Agrell has given the following trenchant characterization of this conventional view in his contribution to a seminar arranged by the 1984 Parliamentary Defence Committee:

> The policy of neutrality is more than most other elements of security policy dependent on the public declarations made in its name. . . . The main purpose of this form of declaratory policy is to function as a means for creating and maintaining the confidence of the world in the credibility of neutrality. This use of doctrinal declarations corresponds both with established practice and the principles emphasized by researchers of doctrinal theory. In the view of the latter, such declarations become *institutionalized* by being repeated often in different contexts and by different representatives and by taking on the imprint of continuity.[78]

He also refers to Katarina Brodin, acute international relations scholar turned Foreign Office mandarin, who has expressed this conventional wisdom well when she notes (in what probably remains the Swedish *locus classicus* on the function of foreign policy doctrines) that 'The problem of credibility has here been viewed as essentially a question of maintaining continuity and consistency in the pursuit of policy'.[79] Agrell goes on to add the following:

> The risk with a declaratory policy of this kind is that its aim of creating consensus and its tendency to contain an open and unbiased discussion of these issues has a negative effect on the ability to analyse the Swedish situation. These declarations easily colour one's view of reality because no alternative view is available for comparative purposes.[80]

In my view, this point is very well taken. Surely, while the credibility of a doctrine *qua* doctrine may quite conceivably depend on such cognitive factors as continuity and consistency (although logical consistency is at least as important), this factor has no decisive bearing on credibility as an essential component of a power relationship. This not only follows from the fact that the latter is a question implicating issues of substantive capacity rather than cognitive effect, but also because a country's power base is both logically and substantively more fundamental than its foreign-policy doctrine. This conception is also closely linked to the implications of the notion of 'intentions' as defined by Jervis. He writes of this concept as follows:

> An actor's intentions can be defined as the actions he will take under given circumstances. . . . It must be emphasized that this definition varies from the common one. Many authors use intention to refer to what the actor plans to do or

what goals he hopes to reach. We are using the term to designate the collection of actions the state will or would take because that is what others are trying to predict.[81]

In this view, credibility is a question of such a prediction of A's intentions by B (in the present context, Sweden's intentions by the two superpowers); and hence it involves not only an analysis of A's past behaviour but also an assessment of what A will do in a future situation. Clearly, consistency in intentions in the conventional sense is only *one* of the components in such a prediction. Much more important is A's *actual capacity* to pursue such given objectives.

In order to avoid some of the more intractable problems raised above, and to elaborate somewhat on the link between vulnerability and credibility posited above, I suggest that we make a fundamental distinction between *three aspects of credibility* in terms of the conceptual logic illustrated in Figure 4.3. The first of these—the *conative* aspect—refers to what Theutenberg has called the 'will' of the neutral state to uphold neutrality. Clearly this is a crucial element in any policy of the kind discussed here: the *manifest intention* of pursuing the policy in question. But as in much of life, 'will' is in itself not sufficient for reaching a desired goal. Rather, the credibility of such a resolve hinges in a crucial sense on it being linked to the *actual pursuit* of the goal in question; and in such a pursuit the prime requisite is the 'capacity'—Theutenberg's second characteristic—for actualizing the endeavour in question. Or as noted by two eminent students of deterrence:

There are two interdependent dimensions of credibility. First, the deterring power must convey to the opponent that it has the *will and resolution* to defend the interest in question; second, it must possess *capabilities* for doing so that it regards—and persuades the opponent to regard—as appropriate and usable for the defense of those interests. The former will have no persuasive force unless the deterring power has the ability to deal, by effective and appropriate measure, with the kinds of action that its opponent may take.[82]

Capacity in this sense is always of a *substantive* nature; and in the present context it is substantive in the sense that we have spoken of the causal conditions underlying all influence patterns between two or more actors. Substantive credibility thus refers to the capacity which an actor in fact can be said to possess for pursuing its expressed intentions, that is for realizing its 'will'. It indicates that the dictum 'Where there's a will, there's a way' is by no means sufficient in international affairs, however credible (i.e. 'believable') the expressed intention or desire underlying it may be. To be credible, a policy must also show that there in fact *is* a way, not only a will. In other words, promissory notes of this kind are effective as foreign-policy instruments only to the extent that they are backed up by sufficient collateral. In foreign affairs, such collateral is quintessentially a question of substantive credibility.

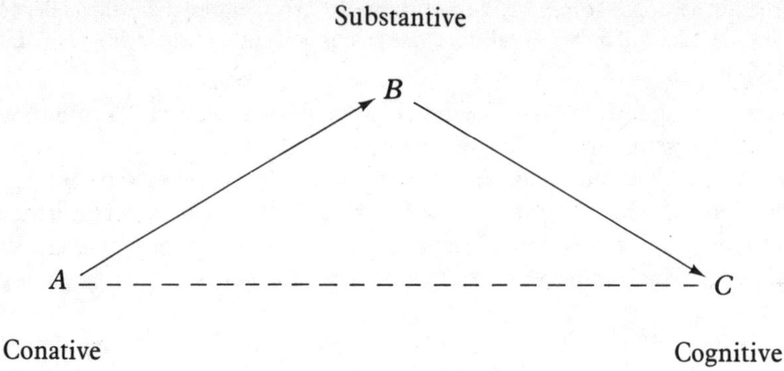

Figure 4.3 The three components of the concept of credibility

Finally, credibility is, if anything, always also a question of *cognition*—of epistemology in action. This aspect is crucial for the following two reasons. First of all, the mere existence of 'will' and 'capacity' on the part of the neutral state is not *per se* sufficient for the purposes discussed here. Knowledge of them must also exist in the minds of the two superpowers.[83] Second, in so far as we are interested in analysing this epistemological factor as an important variable in foreign affairs, we must be able to distinguish between cognitive credibility and its underlying determinants—that which makes it a *variable factor* in the first place. Such an analysis becomes feasible once we conceive of cognitive credibility as a *causal* and *joint function* of the conative and substantive components which, as argued above, necessarily underlie the pursuit of all credible policies. It is also cognitive credibility as conceived here which not only confers that recognition of competence to the neutral actor which, as argued by Ashley, 'empowers' it to act in certain ways but also enables the superpowers to predict—in Jervis's sense of the the notion—the intentions of the neutral state.

From this it also follows that all attempts to create cognitive credibility merely by an act of will—that is by traversing from A to C in Figure 4.3—one commits the *conative fallacy*: the belief, all too common in Swedish (and especially official) discussions of foreign policy, that credibility is merely (or primarily) a question of consistency in expressed intentions, and that such consistency will be the decisive component sustaining both Sweden's 'competence' and 'predictability' as a neutral state. Clearly, principled behaviour of this kind is a necessary but by no means a sufficient condition for achieving credibility in foreign policy.

This is, furthermore, not only a conclusion of fundamental conceptual import in the analysis of Swedish neutrality. Writing on the theme of 'willing what cannot be willed', Elster has suggested that 'Belief, like courage, can be instrumentally useful, and yet be out of reach for instrumental rationality. . . . Beliefs are justified by their ancestors, not by their descendants'.[84] Much the same can be said about credibility, both in private life and in public affairs: one can as little *will* credibility as one can will humility or virtue, however attractive and mutually beneficial these qualities may be to all concerned. The conative fallacy can thus be argued to constitute not only a crucial analytic problem but also a psychological *cul-de-sac* in the pursuit of credibility in foreign affairs. More specifically, and in the terms of the conceptual explication presented here, this means that cognitive credibility—however highly desirable it may be as an anticipated effect: and to a neutral nation it must always remain the crucial composite characteristic of policy—can never be more than a *by-product* of not only the exercise of power, but also, *a fortiori* and ultimately, of the power base of a nation's foreign policy.

This conclusion reiterates the main theme of this discussion as a whole, namely that if it is to be successful, a policy such as Swedish neutrality must of necessity be pursued on the basis of constantly viable power relationships with both of the superpowers, and hence not merely in terms of the public or diplomatic cultivation of positive cognitive effects. Ulysses was well aware of the pre-eminence of this special logic of strategic action. He did not let himself be bound with any other purpose in mind than to make himself less vulnerable—in a very tangible sense—to forces beyond his rational control. It was in so doing, and for no other reason, that this *action*, rather than his *words*, lent both credibility and power to his determination, come what may, to make it safely home to Penelope after his many years of peregrination.

CONCLUSION

The conclusion which I draw from the above analysis is that it is with reference neither to the flow of diplomatic discourse, to public pronouncements nor to doctrinal or political debate that the state of a country's vulnerability to foreign influence or control is to be determined. Vulnerability refers to the power base of foreign policy, specified in terms of the policy-contingency framework in question. It is hence not a question of perception in the first hand, but of actuality (or, in philosophical terms, of ontology rather than epistemology). Although this perspective is not uncommon in the international literature on interdependence and foreign policy, it is not the one which one usually finds in Swedish discussions of her foreign policy, either in scholarly analysis or within the policy-making community.

In the following, concluding chapter such a discussion will be presented with specific reference to the linkage between the two important Swedish policy domains which have been discussed above: energy and foreign policy. I shall there make the argument that the autonomy on which Swedish foreign policy is based is being undermined by energy policies which, despite their domestic responsiveness to issues of energy vulnerability, nevertheless have failed to respond in a manner fully sustaining the credibility of Swedish neutrality.

NOTES

1. Robert Venturi, *Complexity and Contradiction in Architecture* (New York, 1966). Mies van der Rohe was, of course, the eminent architect who created and led the school of *Bauhaus* in Germany and, after Hitler's assumption of power, moved to the United States, where—if we are to believe the various claims of the cognoscenti—his subsequent influence on urban architecture has been both seminal and emasculatory.
2. Wilhelm Agrell, *Bakom ubåtskrisen* (Stockholm, 1986), p. 7.
3. Ibid., pp. 9, 8. My translation.
4. See Jon Elster, *Sour Grapes: Studies in the Subversion of Rationality* (Cambridge, 1983), p. 87.
5. For an analysis of the effects of the 1981 submarine incident in terms of public international law, see Bo Johnson Theutenberg, *Folkrätt och Säkerhetspolitik* (Stockholm, 1986), pp. 431ff.
6. Quoted from *The Odyssey* by Jon Ester, *Ulysses and the Sirens: Studies in Rationality and Irrationality* (Cambridge, 1979), p. 36.
7. See Elster, *Ulysses*, pp. 36ff.
8. See Elster's discussion of the 'globally maximizing machine' in Elster, *Ulysses*, pp. 9ff. See also Thomas C. Schelling's discussion of the related notion of self-command—also demonstrated in terms of Ulysses and the Sirens—in his 'The Intimate Struggle for Self-Command', *Public Interest*, vol. 60, no. 2 (1980), pp. 94ff; and, more recently, the analysis of self-binding commitments in Zeev Maoz and Dan S. Felsenthal, 'Self-Binding Commitments, the Inducement of Trust, Social Choice, and the Theory of International Cooperation', *International Studies Quarterly*, vol. 31 (1987), pp. 177ff, especially pp. 186ff.
9. *Ulysses*, p. 47.
10. For a succinct but informative study of Swedish neutrality prior to the Cold War period, see Krister Wahlbäck's *Den svenska neutralitetens rötter* (Stockholm, 1984). See also Alf W. Johansson and Torbjörn Norman, 'Den svenska neutralitetspolitiken i historisk perspektiv', and, for the period following the First World War, Bo Huldt, 'Svensk nedrustnings- och säkerhetspolitik från tjugotalet till åttiotalet', both in Bo Hugemark (ed.), *Neutralitet och försvar: Perspektiv på svensk säkerhetspolitik 1809–1985* (Stockholm, 1986) pp. 11ff and 185ff. See also the historiography of Swedish neutrality—together with relevant documents—in Nils Andrén, *Power-Balance and Non-Alignment* (Stockholm, 1967).
11. Wahlbäck, pp. 18f. For an English-language history of Swedish neutrality during the First World War, see Steven Koblik, *Sweden: The Neutral Victor* (Lund, 1972). For an international law perspective on the First World War, see Rolf H. Lindholm *Sveriges neutralitet* (Lund, 1987), pp. 40f, 91ff (the former is a summary in English of the latter).
12. Quoted in Wahlbäck, p. 22. My translation.
13. Wahlbäck, p. 24.
14. Agrell, *Spelet om neutraliteten*, p. 13. My translation. Two central analyses of

Swedish policy during the Second World War are Alf Johansson, *Per Albin och kriget* (Stockholm, 1984) and Wilhelm Carlgren, *Svensk utrikespolitik 1939–1945* (Stockholm, 1973).

15. Wilhelm Agrell, 'Behind the Submarine Crisis: Evolution of the Swedish Defence Doctrine and Soviet War Planning', *Cooperation and Conflict*, vol. 21, no. 4 (1986), p. 198.
16. Agrell, *Spelet om neutraliteten*, pp. 13f.
17. Wahlbäck, p. 49. My translation.
18. See ibid., p. 51; and Barbara G. Haskel, *The Scandinavian Option* (Oslo, 1976), p. 41.
19. See Geir Lundestad, *America, Scandinavia, and the Cold War, 1945–1949* (Oslo, 1980); and Haskel, pp. 39ff.
20. Sverker Åström, *Sweden's Policy of Neutrality* (Stockholm, 1983), p. 9. This booklet, published for foreign dissemination by the Swedish Institute, the international information agency of the government, can for all practical purposes be regarded as fully authorized and hence as a semi-official view of Swedish neutrality and foreign policy. Strictly speaking, this official definition means that Sweden's foreign policy should really be referred to in terms of 'non-alignment'—or perhaps even 'not-alignment'—rather than 'neutrality', since the latter pertains only to wartime conditions. However, this is not done in ordinary usage; nor have I found any weighty reasons for departing from this usage here.
21. See Haskel, p. 70; and Agrell, *Alliansfrihet och atombomber*, pp. 383f.
22. Åström, p. 7.
23. Quoted in Wahlbäck, p. 53. My translation.
24. See, for example, Erik Noreen, 'The Nordic Balance: A Security Policy Concept in Theory and Practice', *Cooperation and Conflict*, vol. 18, no. 1 (March 1983), pp. 43ff, and the bibliographic references contained in it; and Gunnar Jervas, 'Sweden in a Less Benign Environment', in Bengt Sundelius (ed.), *The Neutral Democracies and the New Cold War* (Boulder, 1986), pp. 65ff.
25. Åström, p. 9.
26. William Wallace, 'What Price Independence: Sovereignty and Interdependence in British Politics', *International Affairs*, no. 3 (Summer 1986), p. 368.
27. Ibid.
28. Ibid.
29. Quoted in ibid., p. 367. Bengt Sundelius, in a perceptive little essay (unfortunately in Swedish) which goes to the heart of the matter discussed here, has made a similar distinction in terms of the ability to maintain, respectively, 'territorial sovereignty' and 'functional security'. See Bengt Sundelius, 'Neutralitetspolitikens obesvarade frågor', *Internationella Studier*, no. 4 (1984), p. 28.
30. Åström, pp. 6f.
31. The term 'structural constraint' is here used in Steven Lukes's sense of 'precluding (rather than putting a price tag on) . . . a possibility'. Rational constraints, in contrast, *do* involve price tags—which, of course, can render them equally potent *qua* constraints. See Steven Lukes, *Essays in Social Theory* (London, 1977), pp. 12f; and G. John Ikenberry, 'The State and Strategies of International Adjustment', *World Politics*, vol. 39, no. 1 (October 1986), pp. 60ff, where this distinction is applied to the explanation of state preferences.
32. See also Theutenberg, pp. 408ff, who makes a fundamental distinction between the 'doctrine of neutrality' (a legal term) and a 'policy of neutrality', which

constitutes 'that part of our foreign and security policy which has the goal of strengthening, during *peacetime*, our credibility as a neutral state during a war' (pp. 408f; my translation). The Swedish terminology here is, respectively, 'neutralitet' and 'neutralitetspolitik'.

33. Bertrand Russell, *Power* (London, 1975), p. 25.
34. Wilhelm Agrell, *Sveriges civila säkerhet* (Stockholm, 1984), pp. 48f. My translation.
35. Kenneth N. Waltz, in Robert O. Keohane (ed.), *Neorealism and Its Critics* (New York, 1986), p. 90.
36. Agrell, *Sveriges civila säkerhet*, p. 49. My translation.
37. See Karl R. Popper, *Conjectures and Refutations* (London, 1963), pp. 120ff.
38. Felix Oppenheim, *Political Concepts: A Reconstruction* (Oxford, 1981), p. 10.
39. Åström, pp. 8, 9 and 7.
40. Harold Lasswell and Abraham Kaplan, *Power and Society* (New Haven, 1950).
41. David A. Baldwin, 'Power Analysis and World Politics: New Trends versus Old Tendencies', *World Politics*, vol. 31, no. 2 (January 1979), p. 163. My emphasis. It should perhaps be added that Baldwin has—inadvertedly, no doubt—reversed the correct order of these two terms here, in so far as 'domain' refers to the target(s) of influence (his 'whom') and 'scope' to the objectives of the attempt (i.e. to his 'what'). See David A. Baldwin, *Economic Statecraft* (Princeton, 1986), pp. 16f, 20. As also pointed out by Baldwin, the terms as such originate in the Lasswell and Kaplan study previously referred to.
42. Robert A. Dahl, *Modern Political Analysis*, 3rd edn (Englewood Cliffs, 1976), p. 26.
43. See Baldwin, 'Power Analysis', pp. 163ff.
44. See Oppenheim, pp. 10ff. One exception to this rule exists: when the United Nations Security Council decides to impose punitive policies, for example economic sanctions.
45. 'The Polish government', Oppenheim thus writes, 'is unlikely even to consider joining NATO or the Mexican government the Warsaw Pact, even if the Soviet or the United States government, respectively, does not "lift a finger". There is no exercise of power in these cases; yet there is power, and social scientists are especially interested in such latent relationships of interaction, including situations of "anticipated reactions". The more inclusive concept of having power captures such phenomena'. Oppenheim, p. 20.
46. Oppenheim, p. 11.
47. Oppenheim, p. 10.
48. Baldwin, 'Interdependence and Power: A Conceptual Analysis', *International Organization*, vol. 34, no. 4 (Autumn 1980), p. 502. Elsewhere Baldwin adds the following: 'The demonstrations by Schelling that even slight changes in the context of an influence attempt can convert a power asset into a power liability, the recent painful demonstrations that the "most powerful state the world has ever known" could not achieve its goals in Southeast Asia, and the effect of the 1973 Arab oil embargo on U.S. foreign policy suggest that political power resources are much less fungible than has often been implied'. Hence, he concludes, 'In the absence of contrary evidence, I would propose that the international political analyst start with the assumption that power resources are situationally specific. As evidence of fungibility is discovered, appropriate modifications in this working assumption can be made'. Baldwin, 'Power Analysis', p. 168.

49. Lasswell and Kaplan, p. 94.
50. See Jon Elster, *Explaining Technical Change* (Cambridge, 1983), p. 75, and Kjell Goldmann, *Det internationella systemet* (Stockholm, 1978), pp. 55ff. On the role of this type of structuralism in modern world system theories, see Andrew C. Janos, *Politics and Paradigms: Changing Theories of Change in Social Science* (Stanford, 1986), pp. 149ff.
51. Elster, *Explaining Technical Change*, pp. 75f.
52. See also Goldmann, pp. 61ff.
53. For a somewhat different view regarding the 'attitudinal dilemma' of neutrality (not only Swedish), see Bengt Sundelius's discussion in Sundelius, pp. 14ff. For example, on p. 15 he notes the following: 'Whenever the opposing Super Powers stress their distinct value systems, the neutral position becomes somewhat compromised in moral terms. . . . In particular, a sceptical view of the neutral stand is easily justified when such moral concerns are raised by leading members of society in the neutral nations themselves. In this way, the attitudinal dilemma raises the problem of credibility for the chosen security policy. It has important ramifications for the strategic dilemma of the state'. My argument is that although this dilemma may be *experienced* as real, and even be manipulated with such a purpose in mind, it nevertheless remains a spurious dilemma, given the interpretation of the logic of Swedish neutrality presented here.
54. For a brief discussion of the issues involved, see my *Ideology and Foreign Policy; Problems of Comparative Conceptualization* (Oxford, 1986), pp. 176ff. A more extensive analysis can be found in, *inter alia*, Yngve Möller, *Östen Undén: En biografi* (Stockholm, 1986), pp. 291ff, 494ff; and Barry Holmström, 'Realism, moralism och neutralitetspolitik: Östen Undén som anonym ledarskribent', *Statsvetenskaplig tidskrift*, no. 2 (1981), pp. 77ff.
55. Readers proficient in Swedish should in this connection also see Gunnar Sjöstedt's excellent little booklet entitled *Sverige—liten och svagt*? (Stockholm, 1983), especially the conclusion, pp. 29ff.
56. Oppenheim, p. 12.
57. Oppenheim, p. 13.
58. Quoted in ibid.
59. Ibid.
60. Baldwin, 'Power Analysis and World Politics', p. 183.
61. See Baldwin, *Economic Statecraft*, pp. 13ff. Baldwin refers to Harold Lasswell's classic work, *Politics: Who Gets What, When, How*, as the source for this taxonomy.
62. Baldwin makes this assumption. Diplomacy will here, also following him, refer to 'influence attempts relying primarily on negotiation'. Propaganda refers to such attempts relying primarily on the deliberate manipulation of verbal symbols, while economic statecraft relies on 'resources which have a reasonable semblance of a market price in terms of money'. Finally, military statecraft pursues influence in terms of 'violence, weapons, or force'. Baldwin, *Economic Statecraft*, pp. 13f.
63. Incidentally, Baldwin has written of economic instruments of statecraft that they 'may be used to signal narrowly specific intentions or to project general foreign policy orientations. Both are ways to affect the predictions of others about a state's behaviour'. Baldwin, *Economic Statecraft*, p. 102. If we substitute 'super-powers' for 'others' and 'Sweden's' for 'state's', and we think in terms of diplomacy rather than economic statecraft, we here have a very good description

of the type of actions constituting Swedish foreign-policy neutrality.

64. See K.J. Holsti, 'Politics in Command: Foreign Trade as National Security Policy', *International Organization*, vol. 40, no. 3 (Summer 1986), p. 671. On role behaviour see, for example, Stephen G. Walker, 'Role Theory and the Origins of Foreign Policy', in Charles F. Hermann, Charles W. Kegley, Jr, and James N. Rosenau (eds.), *New Directions in the Study of Foreign Policy* (Boston, 1987), pp. 269ff, and the references contained in it.
65. See Elster's discussion in *Ulysses and the Sirens*, pp. 20ff.
66. Baldwin, *Economic Statecraft*, p. 134.
67. Ibid., p. 22.
68. Ibid., p. 23.
69. Richard K. Ashley, 'The Poverty of Neorealism', in Keohane, pp. 291ff.
70. Ibid.
71. Thus, for example, this problem seems intermittently to have reared its head in the otherwise excellent analysis contained in the editor's contribution to Sundelius, pp. 11ff. For an overview of the literature on small states, see Ulf Lindell and Stefan Persson, 'The Paradox of Weak State Power: A Research and Literature Overview', *Cooperation and Conflict*, vol. 21, no. 2 (1986), pp. 79ff; and the papers in O. Höll (ed.), *Small States in Europe and Dependence* (Vienna, 1983).
72. See Baldwin, *Economic Statecraft*, pp. 20ff.
73. See Robert Jervis, *Perception and Misperception in International Politics* (Princeton, 1976), passim. See also Glenn H. Snyder and Paul Diesing, *Conflict Among Nations: Bargaining, Decision Making, and System Structure in International Crises* (Princeton, 1977), pp. 282ff.
74. Two alternatives remain, of course: that an actor is fully cognizant of, and responsive to, its increasing vulnerability to environmental developments detrimental to both its sovereignty and its autonomy; and that an actor is *a priori* unconcerned about potential changes which have not in fact occurred. In this latter case, ignorance is obviously bliss, while the former refers to what is, let us hope, the most common form of the relationship discussed here.
75. Jervas, p. 73.
76. Theutenberg, p. 412. My translation.
77. Ibid., p. 415. My translation.
78. Wilhelm Agrell in *Tio debattinlägg om svensk säkerhetspolitik* (Stockholm, 1985), p. 144. My translation.
79. Katarina Brodin, *Studiet av utrikespolitiska doktriner* (Stockholm, 1977), p. 42. My translation.
80. Agrell, *Tio debattlinlägg*, pp. 144f. My translation.
81. Jervis, p. 48.
82. Gordon A. Craig and Alexander L. George, *Force and Statecraft: Diplomatic Problems of Our Time* (New York, 1983), p. 172.
83. Or as noted analogously of deterrence: 'a sufficiently potent deterring force must exist not only in reality, but more important, in the mind of the potential aggressor as well'. Craig and George, p. 184.
84. Elster, *Sour Grapes*, p. 51.

5 Energy Vulnerability and Swedish National Security

'Since 1973', an American scholar noted a decade ago, 'the United States has had a foreign policy toward energy, even while it has been unable to sort out the domestic consequences of this policy'.[1] In this concluding chapter I shall argue that precisely the opposite has been the case with respect to Sweden during the period discussed here; and that this failure to sort out—and, *a fortiori*, to act in response to—the foreign-policy consequences of her domestic energy policies has arguably had both enduring and undermining effects on Sweden's national security. In so doing, I shall proceed from Richard H. Ullman's conception of the essential dimensions of national security, and especially from his argument that the national security of a state is undermined whenever an action or a sequence of events threatens to diminish drastically and immediately the *quality of life* of its inhabitants, or threatens to narrow significantly the *scope of policy choices* available to governmental and other nationally important decision-makers.[2] In so far as this conception refers to fundamental aspects of economic and political choice and to how such action is circumscribed by international constraints and imperatives, we will thus, ultimately, be concerned with issues of *autonomy* as conceived in the previous chapter, that is with the ability of a state to *exercise influence* over actors and events beyond its borders.

ENERGY VULNERABILITY AND FOREIGN-POLICY AUTONOMY

The Welfare Function of Energy Policy

The first dimension of Ullman's conception of national security—incorporating, broadly speaking, the national welfare function of political action—will in the present context refer to the fundamental societal problems posed by sudden price increases in the wake of oil supply disruptions. As we have seen in Chapter 2, the effects of the two energy crises on the domestic economies and international trade of the oil-dependent states of the West were sudden, traumatic and enduring; and, as we have also seen, Sweden was by no means an exception in this regard, even though—as described in detail in Chapter 3—she was quicker off the mark in managing to legislate comprehensive counter-measures than most other countries hit by the price shocks. In so far as we are here discussing threats to vital national values, in the sense that the vitality of the patterns of economic life decisively determines a nation's ability to pursue and to safeguard the welfare of its population, the question facing us is the following: has Sweden taken out sufficient insurance—in the form of anticipatory energy policies—to counteract the economic consequences of a possible repetition of the first two oil crises? In answering this query, we will not only look at the period and developments analysed in the previous chapters, but also extrapolate from this first decade of oil crises into the future. While so doing, we will keep before us, as a combined *Leitmotif* and measuring rod, the fundamental desiderata of the logic of Swedish neutrality as discussed in the previous chapter. In other words, it is the link between energy policies and the core principles and assumptions underlying Swedish foreign policy that will primarily concern us here, not economic, political or social developments as such.

Vulnerability and the Instability of the Oil Market

When discussing present-day welfare aspects of energy vulnerability, it is necessary as a first step to recognize the inherent instability of the contemporary international oil market. That is to say, if energy vulnerability is conceived as the product of (1) the probability of supply disruptions and of (2) the economic and social costs resulting from inadequate insurance against such disruptions, then—it is here claimed—it would be foolhardy to proceed on the assumption that such vulnerability has decreased (and perhaps even disappeared) since the last oil crisis.[3] Various factors speak against drawing such an unrealistic inference merely from the fact that, contrary to the latter-day prognostications of many energy pundits, we have for some considerable time been blessed with a glut market.

First of all, the Machiavellian principle of circumspection must be given highest priority here, as it must in all contexts involving the national interest. Melvin A. Conant, in a Council on Foreign Relations report in which he notes that 'we know much less about oil—its supply and demand—than most observers would admit', thus writes that 'Prudence compels us to be wary of forecasts that imply that there need be no long-term problem of adequate supply'.[4] Alvin L. Alm, in a Harvard report on American security, draws a parallel in this connection which certainly gives energy policy-makers little cause for taking anything for granted:

> What can be done to minimize this economic misery in future? For most security threats, the antidotes are obvious. For military threats, societies increase armaments and armies. To curb the devastation from floods they build dams, and to prevent outbreaks of disease they inoculate the population. But for energy security there are large differences of opinion about what to do and serious political problems surround-

ing many of the options. Equally important is the fact that most of the implementation steps have not been worked out either as conceptual models or as workable programs. Without having created effective energy security measures in advance, the United States could be saddled with the same flawed mechanisms that failed in the past.[5]

In the face of two major crises within a seven-year period, it would surely constitute an act of gross irresponsibility to ignore or to denigrate the possibility of future repetitions of oil price shocks.

Second, an oversupply of oil on the world market is in itself no guarantee that sudden disruptions are a thing of the past. Thus, although it is difficult to assess the exact effects of declining prices—and hence of declining profits—on the rapidly modernizing countries of the producer regions, such strains are nevertheless generally acknowledged not only to exist but also to constitute a potential threat to the stability of oil supply on the international market. Edward L. Morse thus notes that not only will 'Reduced development efforts—and contraction of the economy— . . . jeopardize the political stability of a number of exporters', but that major industrial countries have already 'organized special efforts to deal with the indebtedness of one oil exporter, Mexico, whose stability and ability to overcome this abrupt change of circumstances is of central foreign policy significance to them'.[6] Hanns Maull, in the same volume on the actual and potential effects of cheaper oil on the international market (appropriately entitled *The Third Oil Shock*), goes one step further when he notes that although the implications of the glut factor for political stability in the producer countries are impossible to quantify, one is nevertheless obliged to 'conclude that politically induced supply disturbances will continue to be quite likely, and that they could happen at any time'.[7]

Third, instability remains an ever-present threat to the world oil market for the simple reason that none of the major conflicts in the Middle East have disappeared since the crisis of 1979–80. Don E. Kash and Robert W. Rycroft, writing about the dangers of American complacency in this regard, suggest why:

The conditions that led to the Arab oil embargo and the revolution in Iran have changed in no fundamental way. An oil disruption could occur tomorrow, next month, next year, or five years hence. It is important to emphasize that global political instability is at the root of oil cut-offs. The nation cannot afford to be sanguine about the availability of imports, even in periods of market surplus.[8]

Two other American scholars have argued similarly—and for similar reasons—in a Brookings report:

Sometime in the coming years it is almost certain that oil supplies will be disrupted as a result of war, revolution, or political action. Most likely, any dangerous disruption will involve the Persian Gulf, since only there are the quantities of oil so large that a major price shock would result.[9]

As implied in this passage, an important reason for the continued vulnerability of the industrialized countries to Gulf oil supply disruptions is that the OPEC countries in this region not only continue to sit on the major repositories of available world reserves, but are also able to disgorge these at a fraction of their cost elsewhere.[10] In short, the crucial problem 'is that cheap, low-cost oil is heavily concentrated in areas with a high degree of political instability, both domestic and international'.[11] As pointed out by a very recent US National Petroleum Council report, the combined effect of these factors may very well mean that by the mid-1990s the international energy market will once again be highly vulnerable to politically motivated actions affecting both the economies and the national security of the oil-importing states.[12]

Fourth, although the oil-importing countries of the West have generally heeded the advice of diversifying their sources of supply in various ways, it nevertheless remains a crucial fact—one that simply will not go away—that, as Joseph N. Nye notes, 'the energy security problem . . . is a geo-political one that centres on the Persian Gulf'.[13] As far as future *price* shocks are concerned, supply diversification is by itself not a viable long-term strategy, given the explosive combination noted above, namely the concentration of the world's most easily accessible and cheapest reserves in one of the world's politically most volatile regions. Diversification is primarily an instrument against the effects of blockades and similar political actions, not against unintended price shocks or manipulations of the oil market for purposes of a quick pecuniary gain.

A fifth factor sustaining the continued instability of the oil market is connected with the glut factor itself, as well as with the inherently cyclical nature of present-day oil market changes. More specifically, I here have in mind the direct effect on demand response of downward movements in price, and especially the constant danger that falling oil prices will undermine the incentive to conserve energy. Although this is not a particularly well-researched topic within energy economics, past experience suggests a strong positive correlation between declining price levels and the increased robustness of demand. Were such a trend to repeat itself in the future, Morse warns, 'the longer-term balance of costs and benefits will shift, and the industrial countries might find their economies even more vulnerable to another disruption than they were to the last'.[14] If governments are inclined at all to undertake actions on the demand side in a situation like this, he adds, they will probably do so for reasons less related to issues of energy security or vulnerability than to budgetary concerns: 'They may judge the costs of any direct form of such intervention to be greater than the long-term losses to energy security of a future, price-induced surge in demand'.[15] A similar warning has been raised by Ulf Lantzke, executive director of the IEA since its creation in 1974. Noting that although the next crisis may in fact never materialize, he adds, however, that the postwar history of the Middle East, combined with the cancellation of projects to find alternative energy sources and a decline in the use of conservation measures, none the less strongly suggest such a possibility. 'It would be a tragedy', he concludes, 'if the events of 1979 and the economic recession that has ensued had to be repeated once again for the lessons to be fully learned'.[16] Furthermore, downward movements in price may also quite conceivably fuel the world economy to such an extent that the present oil supply overhang could be rapidly absorbed, quickly raising demand—and hence prices—once more.[17]

Finally, the politicized nature of large sectors of the present international oil regime continues to remain a destabilizing factor with respect to oil security and price stability. Thus, although the significance of the government-to-government arrangements which characterized the late 1970s and early 1980s seem to have declined as the present decade has progressed, Conant warns us not to overlook 'the point that governments—not the commercial sector—make the big decisions, and that oversupply at any time can become very tight supply in a remarkably short period'.[18] Combined with the various dimensions of the problem discussed here, this factor thus reinforces a situation which, as we have seen above, is anything but stable. This pertains in particular to the supply side of the market, despite numerous claims throughout the 1980s regarding the imminent demise of OPEC. These anticipations seem to have been a function more of wishful thinking than of an objective assessment of its continued influence on central energy issues. Indeed, the resilience of OPEC as the single most dominant actor on the oil market—whether loose or tight—must be regarded as one of the few genuine surprises of latter-day cartel behaviour.

But politics continues to play a crucial role on the demand side as well. Paul Kemezis and Ernest J. Wilson have called this the 'energy policy explosion', describing it as follows:

> This explosion of government intervention has been marked by several factors—much more continuous and high level surveillance of energy issues by senior government officials, more activist policies by government officials at all levels as they are pushed to become more directly involved with domestic and international energy markets, policies aimed for the first time at the intricate linkages between fuels, and in some ways the most radical—the regular involvement of government officials in demand management as well as traditional supply side policies.[19]

As they also make clear, the major reason for this 'explosion' on the policy front was the necessity of governments—on a universal, worldwide basis—to shift from passive to active energy policies following the first oil crisis. This need has not abated, nor are there any indications that it will do so in future.

In view of the inherently unstable market situation described above, it follows that if a nation's welfare—and hence its national security—is not going to become wholly dependent on the ebb and flow of exogenous forces, it requires sufficient insurance against at least the more debilitating effects of sudden price shocks. Essentially, two types of anticipatory policy measures are available for these purposes: *mutual-aid* arrangements between the oil-importing states, and individual *self-help* measures of one kind or another. As we have seen in Chapter 2, the nations in question have in this regard decided to rely on various combinations of both types of measures—albeit in different ways and in terms of different levels of seriousness. More specifically, this has resulted in the creation of the IEA and the various national programmes associated with it. Sweden has been no exception in this regard.

But has Sweden, in so doing, in fact provided herself with the requisite amount of insurance against the adverse national security effects of a new oil crisis? I shall argue below that her reliance on the IEA is based on an over-optimistic assessment of the efficacy of that organization in the event of a new crisis, and that this must lead her to a reconsideration of her policies with respect to the nexus between energy vulnerability and national security.

Vulnerability and the Frailty of the International Energy Agency

As we have noted earlier in the present study, the viability of the IEA as an energy insurance regime has been questioned ever since its inception following the first oil crisis. In my view, the following criticisms against the crisis management utility of this organization are both persuasive as analytic arguments and empirically relevant to our particular concerns. As such they also highlight some of the fundamental weaknesses not only of the foundations supporting present-day Swedish energy policies but also, *a fortiori*, of her national security policies more broadly conceived.

First of all, it must be kept in mind that the impetus for creating the IEA was primarily political rather than economic. As its director-general has pointed out, what the 1973–4 oil embargo brought home to the oil-importing countries was in the first place a recognition of 'the political vulnerability associated with import dependence'.[20] This has meant, amongst other things, that the workability of the IEP has from the very beginning been closely linked to the *political will* of the IEA's member states—especially to the political will of its most powerful members (who also happen, with a few notable exceptions, to be oil producers themselves).

In the immediate aftermath of the embargo, this will was successfully harnessed as a consequence of Henry Kissinger's diplomatic adroitness, backed up (when this was found necessary—which it was) by appropriate arm-twisting on the part of President Nixon. However, this genesis of the organization in Washington has not only led OPEC to regard it as an essentially confrontational body—a stigmatization which it has not really managed to rid itself of since then—but also left it a legacy which has been largely counterproductive in the subsequent history of the organization. As we have seen in a previous chapter, this legacy is connected with prevailing assumptions that the IEA was created either to consolidate US relations with Western Europe and Japan in order to prevent direct links between the latter and the OPEC countries; to defend the signatories against one another in the event of an energy crisis; or simply to defend American interests in general.[21]

Furthermore, as noted by John Ikenberry, the United States quickly became disillusioned with the IEA as an effective instrument for pursuing an international offensive strategy to curtail the price-setting ability of OPEC. Instead, she was forced increasingly to rely on domestic measures. However, to their chagrin both the Ford and the Carter administrations soon found out that domestic American political institutions were simply not up to this task: 'International offensive strategy needs to be undergirded by sufficient and relevant domestic resources and capacities. In this case, the capacity to reduce oil imports—and thus take the upward pressure off international prices—was missing'.[22] Given both the circumstances of the organization's inception and these dispiriting experiences on the part of its major initiator, it should not come entirely as a surprise that the political credibility of the IEA has, on the whole, never been particularly high. Or as Alm has put it:

> There is little basis for consensus among IEA members on the issue of international cooperation in energy security. Many Americans do not believe the IEA agreement will actually work, and yet it continues to remain a legal commitment among IEA participants. Many otherwise knowledgeable people are concerned that oil will be shipped from the United States to other countries as a result of the IEA agreement—an eventuality that is unlikely. US policymakers have been wary of schemes to promote cooperation in managing stock withdrawals, even though the United States, which will have the largest reserve, has much to gain.[23]

Clearly, this type of widespread scepticism and the lack of a political will which it betrays raises very serious doubts about the ability of the IEA to harness its member states in the event of a new crisis. This is especially true if, as Edward N. Krapels has surmised, 'Importers that are "hostages" in a confrontation in which they have no strong political interest are likely to be less willing to endure economic losses or to deplete their emergency reserves for the sake of the foreign policies of other importers'.[24] Indeed, even if the automaticity of the IEP may cause agency members to enter into a crisis collectively, subsequent developments—especially in view of the risks involved in rapid stock depletion—may very well cause some importers to abandon the programme subsequently.[25]

Second, the issues of credibility and co-operation involved here can also be discussed more generally in terms of the substantive implications of two types of theoretical 'games' which have often been used to analyse problems of this kind. Thus one can, first of all, construe the IEA agreement as an attempt to create a public good premised on the collective provision of a non-excludable and joint benefit.[26] Clearly such an aim was central to this whole venture: to provide for the continued availability to all of oil during supply shortages, especially shortages caused by purposive behaviour on the part of the producers. By assuring that oil would be made available to the consumers of the industrialized West at affordable prices even in the face of new blockades of the 1973 type, it was hoped not only that the international economy would be left largely unaffected during future shortfalls, but also that the producers would be made to realize the futility of repeating such punitive actions and hence refrain from them altogether. Thus while the thrust of the provisions of the IEP was directed at a finite number of oil producers, the intended benefits accruing from them were generally conceived of in collective good terms, that is in terms of the international oil market itself *qua* open trade system.

However, a major problem with collective goods of this kind is, as Russell Hardin has phrased it, that 'the costliness or de facto infeasibility of exclusion from consumption of a collectively provided good usually eliminates any direct incentive for individual consumers to pay for the good'.[27] In other words, the gist of this theory is the claim that there is a built-in tendency towards *free-riding* on the part of participants involved in interest-group activity created for purposes of the collective provision of non-excludable goods. From this it also follows that the pre-eminent problem in a collective goods situation is, in essence, the prevention of this type of exclusively self-interested behaviour. More specifically, the problem here is that groups or organizations founded spontaneously as a consequence of single events or crises will tend to dissolve once the occasion—that is the catalytic event itself—is over. In such instances, Hardin explains, 'the individuals' activities are no longer coordinated, and the group returns to its latent state even though its interest may not yet have been secured'.[28]

Various strategies have been proposed to deal with this problem. In a discussion of public goods and the international political economy, John Conybeare, for example, has argued that the appropriate response must essentially be based on some form of 'compellence' (he here refers to Thomas C. Schelling's classic discussion), an approach aimed at persuading actors to do what they would not be inclined to do if left to their own devices.[29] However, he immediately adds that such a strategy is fraught with difficulties:

> Compellence threats to induce contributions to a public good must actively specify the act required and the time permitted for it to be performed, while at the same time conveying assurances that there will be a payoff for cooperation and that the compellant threat could easily be resumed if cooperation is not forthcoming. All this must be done while taking care not to make compliance with the compellent threat too conspicuous.[30]

More generally, the problem here is that compellence involves bargaining in the form of 'negative coercion'; and clearly, if it is to work at all, such a strategy calls for a strong protagonist (or hegemon) if the free-rider, or a group of free-riders, is to be persuaded to produce the appropriate contribution to the provision of the public good.[31] Mancur Olson, on the other hand, takes a considerably less strong-armed approach in his claim for the efficacy—at least sometimes—of selective incentives rather than coercion, that is incentives based on private rather than collective good interests. The main problem with such incentives is, however, that generally they are not sufficiently powerful to override the attraction of free-riding. A third and related strategy is to hope for the emergence of appropriate political entrepreneurs, that is people who find it in their private interest to work for the provision of collective benefits. This strategy has also been criticized for various reasons—mainly that while the existence of such entrepreneurs tends to explain *successful* collective action *ex post facto*, it has not shown itself to be amenable to effective implementation in the face of floundering public goods associations.[32]

Both the logic of free-riding and the problematic nature of the strategies suggested above for overcoming it are, in my view, highly relevant to an assessment of the IEA's utility in energy crisis management. Thus, first of all, we here have a 'rational' explanation for one of the most intractable problems which has plagued the organization since its earliest days: the presence—indeed, omnipresence—of free-riding activity in the form of widespread nonchalance regarding the maintenance of sufficient stocks and effective restraint measures and, especially, the implementation of oil import reduction policies (conservation in particular). Second, the record considered in the first chapters of this book is clear on the absence of compellence as a strategy for inhibiting these concrete manifestations of the logic of free-riding. Nor should this surprise us, given the fact that the type of compellence which is at issue here requires a collective international decision sufficiently authoritative to override all the individual national interests of the member states involved. Since the United States was no longer hegemonic in international energy policy during the period in question, and has in any case at best been a lackadaisical member of the IEA during most of its history, who could in fact have provided the necessary wherewithal for making compellence a viable strategy? Certainly not the IEA itself.[33] Nor has it been possible to discern much activity during the decade in question in the form of political—as distinguished from pecuniarily more remunerative—forms of entrepreneurship on the international energy arena.

All in all, one is tempted to conclude that the reason the IEA still exists today as an ongoing international concern is not a function of its crisis management propensities; it still exists because of the selective incentive to its members which is provided by the organization's role as an—apparently efficient and certainly a cost-effective—instrument for collecting and disseminating information on the worldwide energy situation. Although this is indubitably a positive function, in so far as it provides an empirical base for potentially sound national energy policies, it is nevertheless not one which carries much clout in the face of the issues discussed here.

Although free-riding is an effect of the non-excludability aspect of collective goods, such goods are usually also defined in terms of a second quality, namely joint supply, a notion referring to the existence of non-rivalry in consumption. The consequence of this second factor is 'that the existence of free-riding does not alter the incentives for other contributors to change their own provision of the public good'.[34] In other words, the fact of free-riding does not as such imply that it would be disadvantageous *per se* for the non-free-riders to continue paying toward the purchase of the collective good. Rather, as Conybeare has noted, the pertinent logic here is the following:

> Where there is a small number of beneficiaries, suboptimal supply will result from each supplying the public good up to where marginal costs equal private marginal benefits, and because there may be strategic behaviour amongst the beneficiaries (for example, not revealing your preference for the good so that others will provide it, or playing 'chicken' by threatening to make no contribution). As the number of beneficiaries rises, the optimal contribution for each beneficiary tends to zero (due to falling marginal benefit) under most conjectural variations on the model, such that the problem acquires the payoff characteristic of a prisoners' dilemma game.[35]

This brings us to the second game which, in my view, is relevant in an analysis of the viability of the IEA in the event of a new oil crisis: that of the Prisoners' Dilemma.

The argument here is that although nothing is to be gained in a public-goods game from attempting to exclude an actor from the aims of the game as long as the non-rivalry characteristic is present (which, in theory, it is as long as IEA remains a practicable insurance regime for keeping oil flowing at an acceptable price during a crisis), the situation changes as the number of free-riders increases and—partly as a consequence of this—jointness of supply itself becomes the crucial issue: which is obviously the case during an oil crisis. As mentioned in passing in the first chapter, when this occurs the logic of the Prisoners' Dilemma enters into play, and with it the 'beggar thy neighbour' policies which—since we have dealt with them in some detail above—need not detain us here.

According to the logic of this second game, the problem is to dissuade all players from taking action aimed only at satisfying individual interests. Referring to Schelling's terminology once more, the pertinent logical strategy here is the obverse of compellence, namely deterrence, an action intended to persuade an actor to refrain from doing something which he or she would otherwise be inclined to do.[36] However, although potential strategies for overcoming the suboptimal consequences of the Prisoners' Dilemma have been widely discussed, the facts of the present case do not give us much to be hopeful about on this score either, especially if, as I have suggested above, this situation is essentially a structural consequence of free-riding and, more generally, of the deep-rooted problems of collective action in international politics.[37]

My conclusion here is, in short, that in so far as the IEP is premised on collective action in the persuance of a public good, and as long as this is done without recourse to a hegemonic state able to exercise both compellence and deterrence when necessary, the logic of the situation will tend to favour the pursuit of individual over collective interests. As we have seen in a previous chapter, this is also what has happened in reality. Or as two knowledgeable energy analysts have noted, once the crunch set in during the spring of 1979, the governments of the oil-consuming countries 'abandoned the policy of benign neglect which they had so far pursued, for one of "sauve qui peut"'.[38] This passage expresses our problem in a nutshell: first free-riding, then each one for himself.

This discussion of the potential role of the IEA in the event of a new oil crisis will be rounded off by referring—with little editorial comment—to a highly pertinent instance from real life. I shall quote at some length from Keohane's version of the incident in question, which is based not only on the well-informed account offered by Daniel Badger and Robert Belgrave (the former an IEA official, the latter well connected both with the oil industry and the British government), but also on internal IEA documents as well as on extensive discussions with IEA officials in 1981.[39] Keohane writes the following about some crucial events during the 1979 oil crisis:

> In February of that year Sweden, which maintained a system of oil price controls, complained that it was suffering a 17 per cent shortfall of supplies for the first quarter. The secretariat's reaction was to argue that a nominal shortfall was normal for Sweden during the winter, owing to seasonal variations. Because ports in northern Europe become blocked by ice, these countries normally build up stocks in the summer and draw them down during the winter. From February to May 1979 the secretariat negotiated with the Swedish government to adjust its calculations of supply patterns, in an attempt to avoid a formal request that the emergency system be triggered. Finally, in May 1979, Sweden did call for activation of the emergency system, noting a reduction in first-quarter supply of 9.9 per cent, after adjusting for seasonal factors. Four days later the governing board declined to accept this request.[40]

Why did the governing board of the IEA not follow the rules of the IEP, since we here had an obvious case of a shortfall above the level—7 per cent—calling for an 'automatic' activation of emergency measures? Because, Keohane notes, referring to the secretariat's reasoning, 'such an action would have been a huge gamble. Hoarding and panic could have been the initial reaction, worsening the situation. . . . If the system did not work, confidence in the IEA's emergency system as a response to large-scale disruptions could have been lost, and the IEA itself might have been in danger'.[41] The most powerful members of the agency agreed with this conclusion. Consequently, for 'the rest of the year the IEA relied on informal consultations . . . rather than seriously considering formal implementation of either the selective or the general sharing arrangements'. Keohane's conclusion is that instead of constituting meaningful guarantees, 'IEA rules served as symbols, used by the custodians of the regime to legitimize their behind-the-scenes attempts to facilitate agreements'. To do otherwise 'could have threatened the organization's existence as well as its already limited utility to inspire confidence in the efficacy of its actions and the perspicacity of its judgment'.[42] The point hardly needs to be rubbed in that what we have here is not much of an insurance policy—especially not if a state is small, heavily dependent on oil for its energy needs and without any wells of its own.

Before moving on to the next and final section of this study, let me here conclude by anticipating a few objections to the line of reasoning pursued above. The gist of the argument here is the claim not only that the present oil market is considerably more unstable than is often assumed, but also that Swedish energy policy has serious weaknesses as a consequence of its heavy reliance on the IEA as an insurance regime against the effects of future price shocks. Taken together these two factors should, in my view, make all pertinent decision-makers (as well as foreign-policy scholars and professional energy analysts) considerably more circumspect in their assessment of Sweden's ability to withstand a new energy crisis. However, as against this general indictment of present policies, one can of course argue that Swedish energy policy is by no means based solely on the viability of the IEP or on other activities—formal or informal—pursued within the aegis of the IEA. After all, as noted in such painstaking detail in Chapter 3, Sweden has undoubtedly made a successful effort to reduce not only her dependence on imported oil but also her sensitivity to market disruptions in a more general sense. In short, it can be claimed that given the robust state of Sweden's present energy policy, her national welfare is relatively safe from the possible ravages of a new crisis, irrespective of IEA provisions and responses. As against this cheery optimism, the following caveats and reminders are pertinent.

First of all, the claim that Sweden is better prepared against a crisis today than in the late 1970s does not, even if true, necessarily imply that she is therefore *sufficiently* insured against the negative effects of future price shocks. One major drawback with successfully implemented energy policies—especially policies of conservation—is that the margin for further reductions in demand is thereby *pari passu* reduced, leaving less in the way of a fall-back position. This is, indeed, one of the reasons for the occasional apprehension that the effects of a third oil crisis will be even worse than those following the first two: there will simply be far less fat to trim this time around. In such a situation the importance of sufficient reserves increases—that is stocks available for an immediate alleviation of domestic problems of supply and price. And on this front, quite clearly, Sweden's fortunes are still in substance intimately linked to those of the IEA, since her peacetime stockpile levels have constantly been pegged—rhetorically, nominally or in actual fact—not only to the ninety-day stock requirement stipulated by the IEP but also to the emergency sharing provisions to which she is a signatory. Although the problematic nature of this double dependence—evaluational as well as structural—on the IEA has been adumbrated in official Swedish reports, the fact remains that at no point has a serious analysis been made to determine the sufficiency of this level as far as Sweden is concerned; nor if, in a crunch, it will indeed constitute a practicable instrument for domestic price stabilization. This situation is further exacerbated by the following facts:

> No system exists to draw down stocks apart from the IEA sharing agreement. Under the agreement, stock withdrawal is assumed to meet any country's shortfall that is not met by conservation. The ninety-day stock requirement is not aimed at reducing price impacts from an interruption, does not contain provisions for cooperation among the participating countries, and makes no provision whatsoever for drawing down stocks during interruptions of 7 per cent or less.[43]

Indeed, given the logic of a protracted crisis, it has even been questioned if decision-makers facing such a situation will be willing at all—for national security reasons—to take the risk of using more than small amounts of such stocks for purposes of minimizing the welfare effects of price increases. Clearly, in a situation of this kind one of the most widely accepted laws in social science comes into play, namely the law of diminishing marginal utility. This law states that, other things being equal, the less of something one has, the higher the value of a single unit (and, of course, the more one has of it, the lower the value of a single unit).[44] In other words, the more a nation's energy stocks decline in volume, the less inclined policy-makers will be to use what quickly becomes an increasingly scarce strategic commodity.[45] This law is especially pertinent in the event of a crisis combining supply disruptions with other types of threats to a state's national security. The policy-maker's ultimate nightmare is to be sucked into a severe crisis with no stocks left at all.

The problems discussed here must also be viewed against the backdrop of the broader political-economic framework characterizing Sweden's present-day international situation. In a widely discussed essay on this topic by Gunnar Sjöstedt, a forceful case has been made for the claim that Sweden's relatively strong postwar international standing has declined during the past two decades, primarily as a consequence of a diminishing domestic economic base coupled with her increasing beholdenness to the larger international economic system. As noted by the above author, these developments sustaining Sweden's increasing vulnerability to exogenous factors not only tend to have a negative effect on her general economic welfare but have already, in his view, undermined the preconditions for Swedish neutrality itself.[46]

Finally, it is also important in this context to reiterate the point made by Morse above, namely that given the prevailing oil market, the long-term balance of costs and benefits will probably shift once more, with results that might even be worse than those following the previous crises.[47] He gives the following grounds for this projection:

Governments will be inclined to undertake actions on the demand side, in response to lower prices, but probably for reasons related less to energy security than to budgets. Governments in consuming countries are likely to see the energy sector more as a source of revenue than as one related to the fundamental welfare of society as a whole.[48]

Such a development is already under way in Sweden, in the form of what must be considered to be one of the most myopic policies ever conceived in the context of the present-day energy situation: the decision, in the middle of an eroding oil market, to sell off large chunks of peacetime reserves mainly for the purpose of ameliorating short-term strains on the national budget.[49] It is probably due to the thinking underlying this type of action that Wilhelm Agrell was prompted to make what is undoubtedly a harsh but, given the arguments above, nevertheless a fair summation of the decade which we have considered here: 'Although ten years have passed since the first oil crisis, Sweden has not really even started in any fundamental sense to solve the problems of energy vulnerability'.[50]

Choices, Constraints and Energy Policy

In its negative form, the welfare function of energy policy has here referred to the societal effects caused by sudden price increases following oil supply disruptions. I have argued that in so far as Sweden has not sufficiently insured herself against the more egregious consequences of potential future oil crises, such disruptions could substantially diminish the welfare of the nation, thus implicating issues of national security in the sense conceived by Ullman.

However, as previously noted, Ullman has also suggested that challenges to national security can, in addition, be defined in terms of external conditions and actions which threaten to narrow significantly the scope of policy choice available to a government. It is to this aspect of the link between energy vulnerability and Swedish neutrality that we shall now turn. Although it is obvious that the welfare issues discussed above are directly relevant to Sweden's capability of pursuing her foreign policy effectively, it is only when the former is conjoined with an analysis of the latter that the core dimensions of this policy can be fully evaluated. In other words, we will now consider the claim that Sweden's present energy vulnerability is not only a question of domestic welfare and economic clout but, in the final analysis, an issue which affects the *sovereignty* and, above all, the *autonomy* of Sweden in the present-day international system. In so far as this is the case, this factor has consequences for Sweden's *exercise of influence* vis-à-vis the two superpowers, and hence for the policy of neutrality itself.

Vulnerability and National Sovereignty

When discussing the notion of precommitment in the previous chapter, it was suggested that the notion of sovereignty is crucial for understanding the reason for Sweden's adamant refusal to assume international obligations which might be construed to impugn her sovereign ability to make independent decisions on national security issues. Retaining for herself this sovereign right lies at the core of the policy of neutrality, since it both substantiates and epitomizes Sweden's commitment to remain wholly free of all formal external ties in the pursuit of her national security.

It is in terms of this argument, for example, that Sweden—after years of vacillation—finally concluded in March 1971 that membership in the EEC was not a realistic option on her part. Participation in an integrative process which had the aim of eventually consolidating into an economic and monetary union was not deemed compatible with Swedish neutrality, since it 'would involve the giving up of the national right to make own decisions in important fields'.[51] The problem with joining the EEC was thus not primarily one of becoming too dependent on the EEC (or its member countries), but of having to forfeit certain sovereign rights. In so far as Sweden's neutrality is premised *inter alia* on the presumption that she must at all times remain free to make an independent determination of her foreign policy on every important issue, taking this step was obviously not possible without giving up neutrality itself. This is, however, a conclusion which follows from Sweden's own specific definition of her neutrality, not from any innate qualities inhering in either sovereignty or neutrality as defined by international jurisprudence. Indeed, many countries—for example in Sweden's immediate vicinity, both Norway and Denmark—have purposely given up certain sovereign rights in order to enhance rather than to diminish their national security. So, in her special way, has Finland, a neutral country. Thus the fact that Sweden has decided not to join her neighbours in this respect emphasizes, if anything, the importance of this issue when assessing the effects of energy vulnerability on her foreign policy.

In the previous section of this chapter my main line of argumentation was presented in the form of two general propositions: that the international oil market has remained inherently unstable (and will continue to do so in the forseeable future), which means that a new crisis can potentially occur at any time; and that the IEA is too frail to act as an effective crisis manager in the case of such an event. These two factors can be said to define the main exogenous sources of a nation's energy vulnerability today. In the present section the thrust of my argument may at first seem paradoxical, since I will now try to make a case for the claim that *even if* the IEP were a fully viable instrument against the effects of a new oil crisis, this fact would nevertheless not help to assuage the negative consequences for Sweden's national security of her present energy policies and the vulnerability underlying them. However, it must immediately be added that whereas the arguments presented in the previous section have been posited as valid *prima facia* for all industrialized states dependent on imported oil, the arguments below will pertain only to the Swedish instance. In other words, I will now discuss energy vulnerability exclusively in terms of its consequences within the specific framework constituted by the logic of Swedish neutrality. That is to say, I will now proceed to show that given this logic, not just any energy insurance regime—however viable it may be as such—will do as far as Swedish national security is concerned.

The crucial aspect of the IEP with respect to the issue of national sovereignty is that it poses clear and substantial obligations on the participants of the programme. In this connection Louis Turner, one of Great Britain's most knowledgeable energy analyst, has commented on the main provisions of the IEP as follows:

> These measures were far more than vague statements of desired actions in times of crisis. They were, rather, obligations under international agreements involving member countries *in a surrender of sovereignty in some energy matters. Given the importance of oil and other sources of energy to national well-being, the nearest comparison one can make is with the obligations incurred by members of NATO*.[52]

The main reason why the IEP has the consequences indicated by Turner is that the unanimity rule is not applied in IEA decision-making (in distinction, for example, to the OECD). This means that it is very difficult for a single country to block, or to exempt itself from, the implementation of joint IEA measures in a crisis situation. This voting structure gives the agency whatever strength it may have as a supranational organization, but at the same time it has the implication, for example, that it makes it next to impossible for the West Europeans, including Sweden, to refuse to help the United States if the latter were once again hit by a selective embargo.[53] They would be obligated to do this in the event that the Secretariat of the IEA makes the proper recommendation and its Governing Board does not block it by means of a special majority.

Upon gaining membership of the IEA, Sweden insisted on appending a conditional declaration to it regarding her neutrality. It is significant to note, however, that she did not choose the route taken by Norway, the only country which insisted on negotiating a special relationship with the IEA because of the nature and extent of the obligations involved. As Turner has written, the latter 'was extremely reluctant to join an organization which might appear to limit her sovereignty over her off-shore oil. It needed a special dispensation, formally excepting Norway from obligations in any crisis, before she was willing to participate in the rest of the IEA's activities'.[54] The reason why Sweden did not proceed similarly is not difficult to understand: she simply did not have either the same interests or bargaining power, being both extremely dependent on imported oil for her overall energy needs and possessing not a single well of her own. At the same time,

Olof Palme told Parliament in 1975 that by joining the IEA in the way that she did, Sweden was 'assured the same access to oil supplies as the other members of the organization. Of special importance is that the distribution of oil [during a crisis] is effectuated according to predetermined rules'—the rules alluded to above.[55] This would seem to indicate that Sweden could have it both ways: joining the IEA without compromising her neutrality. This is also the way her membership has been interpreted officially.

I would like to argue, however, that given the whole point of Sweden's desire to join the IEA—the need to vouchsafe access to oil supplies during interruptions—it is very difficult not to conclude that for all practical purposes the declaration of neutrality appended to the agreement does not diminish or change Sweden's actual obligations to the organization, that is those obligations which Turner has described in terms of a partial surrender of sovereignty. Indeed, one of the very crucial aspects of this loss of freedom is contained in the very stipulation pointed to by Palme above: the redistribution of oil during a crisis according to predetermined rules. It is precisely these rules which enable the IEA as a supranational body to override national objections during a crisis—an advantage to most members, especially the weaker ones (which is probably what Palme had in mind), but at the same time necessarily implicating a breach of national sovereignty. In any case, since it is if anything during such a crisis that Sweden would be in most need of IEA's collective provisions, this would be the least propitious time for Sweden to invoke her special status as a neutral state.

It is not clear whether Sweden assumed these IEA obligations with open eyes, or whether the implications of her membership which I have suggested here were not fully recognized at the time of her accession to the programme. However, it is of some interest in the present context to note that during the very same period, 1974–5, Sweden found herself in another situation which could also be construed to involve a potential erosion of her neutrality. I here have in mind the attempt by the Swedish aircraft manufacturer SAAB to sell the Viggen aircraft to four NATO members (Norway, Denmark, Belgium and the Netherlands), and the Swedish government's commitment to these states that it would guarantee the continued delivery—even in wartime—of those parts to the Viggen which would be produced in Sweden.[56] Clearly, had the Swedish government actually gone ahead with this commitment in the event that SAAB had won the contract, this would not only have constituted a breach of neutrality according to international law but would also 'have meant that the credibility of the Swedish policy of neutrality would have been lost', as a Swedish diplomat and scholar of international law has recently noted.[57] He has also raised the question why the Swedish government decided to take this risk at all (which in itself had clear negative implications at the time with respect to at least the conative credibility of Swedish neutrality), especially since the chances that SAAB would actually land the contract were so slim. His conclusion is that there 'is no clear-cut answer to this question so far', but that 'for the sake of the Swedish policy of neutrality

. . . it must be looked upon as fortunate that the Viggen venture failed'.[58] Although the Viggen case is obviously both more serious and clear-cut, the same conclusion must be said to apply to Sweden's IEA membership; and hence that it is fortunate that the IEP emergency provisions have never been activated. This has, however, not prevented open criticism on both scores by at least one neighbouring (and neutral) state. Nils Andrén has thus noted that 'in Finland criticism is sometimes voiced against Sweden for casting doubts on its own policy of neutrality, e.g., by joining the IEA or by trying to market the Viggen aircraft in NATO countries'.[59] Obviously, if such doubts are raised by what is not only a friendly—as well as co-neutral—neighbour but also one with a special relationship with one of the superpowers, they should be taken seriously.

Vulnerability and National Autonomy

More serious in the present context is the final aspect of the relationship between Sweden's energy vulnerability and her neutrality which I intend to consider here: the ability to retain her autonomy in the event of a new oil crisis. If sovereignty implicates issues of 'will' and conative credibility, autonomy refers to 'capacity' and hence to the substantive credibility of Swedish neutrality. Since I have discussed the nature and importance of these notions in detail in the previous chapter, I need not repeat that discussion here. Instead I shall briefly indicate why, even if the IEP did work as envisaged by its most optimistic proponents, this would arguably not decrease the negative effects of Sweden's present energy vulnerability on her autonomy as a neutral state. The gist of this argument is that in so far as it is believed by many that the IEA is essentially an agency created to protect the United States and her allies from OPEC malefaction or greed, it can be claimed to be *de facto* part and parcel of the very system—that of the bipolar world of East–West confrontation—from which the logic of neutrality requires strict Swedish disassociation at all times. As long as Sweden is incapable, wittingly or not, of maintaining such autonomy, the credibility of her neutrality is in constant danger of being challenged and undermined.

The problem with the whole idea of the IEA from the point of view of a neutral state is not only its beholdenness to Washington, but to a Washington which is not overly fastidious about upholding any fundamental distinctions between 'high' and 'low' politics. Thus, although the first energy crisis was in most capitals of the industrialized world treated as essentially an economic problem (albeit with certain political implications), the American view was quite different. This difference is nicely set down in an interview which Henry Kissinger gave *Business Week* in January 1975, in which he noted that 'you could never implement the energy policy as a purely economic matter. It has been a foreign policy matter since the beginning'.[60] Although this view was subsequently more or less turned on its head by the first Reagan administration (in the process almost killing off the IEA as well), it suffused the policy and practice of the Nixon, Ford and Carter eras and could quite conceivably resurface again in the event of a new crisis.

The underlying reason for this American view is in fact not difficult to understand, given the crucial importance of oil in all industrialized countries. Or as Conant has put it:

> The fundamental requirement remains: oil is a strategic commodity; its availability and on what terms is as much a part of collective security and mutual support as is NATO and the US security relationship with Japan. The energy vulnerabilities of the allies affect the overall strategic balance with the USSR. For exactly the same reason that the allies have found it necessary for their own survival to link their defences—individually, none could survive Soviet pressure or attack—so they need to combine their economic, financial, and political assets to help ensure their societies survive some failure in oil supply.[61]

But if this view is readily comprehensible, and hence also the need for NATO countries to co-ordinate not only their military but also their energy policies, it should be equally transparent (as Krister Wahlbäck has contended) that 'Sweden and Finland as neutrals have to consider carefully the extent to which they are interdependent with big NATO countries. They have to make credible their capacity to withstand Western political pressure by economic and other means in time of crisis or war'.[62] Although Wahlbäck is here thinking mainly in terms of the consequences of such interdependence during crisis periods, the warning raised by him is, in my view, equally valid with respect to the issue of substantive credibility as such, given a self-imposed logic which requires that even in peacetime Sweden must continuously demonstrate her capacity to act as an autonomous agent in all matters of national security.

But if this is so obviously the case, why has this issue not become more central in Swedish discussions of energy policy? Since this is a complex question implicating sticky issues of causality, I will here venture only a few suggestions. One reason is that whereas the American approach to energy policy is characterized by 'the kind of latitude associated with military policy', the Europeans have in general viewed such policy 'as a brand of industrial policy to be managed as well as possible given major difficulties'. The goal of this approach, Kemezis and Wilson add, 'is not to "win" the energy war, but rather to make whatever national adjustments are possible to ease problems and maintain a minimum of national security'.[63] Sweden seems willy nilly to have been pulled along by this European viewpoint, even though it should have been obvious to her that while most European states could pursue this low-key route in the confident knowledge that in a real crunch the United States would step in with her 'military' approach, this fall-back position would not be compatible with Swedish neutrality.

This slide into a compromising situation can, in turn, be blamed to a considerable extent on the extraordinary intensity of the nuclear energy debate during most of the decade considered here. Domestic politics in the form of this issue were, it seems, allowed to trump foreign-policy considerations on every turn. As shown in Chapter 3, this domestic issue not only 'neutralized' the traditional left–right dimension of politics but also destabilized the whole political arena itself for a long time. The ecology issue seemed to have become so important that it effectively squashed any serious consideration of its effects on foreign policy. In this sense it substantiates very well the following conclusion posited by Leon Lindberg:

> Energy policy choices, although severely constrained by international factors, are overwhelmingly determined by the push and shove of internal political life with very limited consideration of international consequences that may have new and more dangerous influences on subsequent policy . . . most governments seek where possible to lower the costs of national adaptation to the changing energy situation by directly or indirectly 'externalizing' these costs.[64]

In all probability this pervasive concern with the environmental and safety effects of nuclear energy also played a central role in determining the thrust of Swedish research and development efforts, especially after the accession of the Centre Party to the premiership in 1976. If so, it also goes a long way in explaining why, despite the immense investments on R & D during these years, not a single serious study of the relationship between energy policy and Swedish national security saw public daylight. This noteworthy lacuna during a decade of great energy policy rejuvenation is, in my view, typified by the paucity of Swedish stockpile ambitions: to reach the minimal target set by the IEA within a ten-year period.

Finally, it must also be acknowledged that if, as David A. Baldwin has written, 'the opportunity costs of autonomy are prohibitively high', the same applies *pari passu* to Swedish neutrality.[65] In such a situation the temptation of simply adhering to the 'politics of the acceptable'—that is, giving way to the exingencies of 'adaptation' and 'drift' in policy implementation—becomes particularly acute. In this regard the Swedish public and its decision makers have probably had a view of neutrality which is too optimistic about its 'acceptable' limits and underlying capacity.[66] If nothing else, the so-called submarine crisis should have put paid to such assumptions of foreign-policy latitude.

Whatever the ultimate reasons for this development in energy policy, the conclusion which lies close at hand here is that in so far as her energy policy is tied to American foreign-policy considerations in the sense adumbrated above, there would appear to exist *prima facie* grounds, on the part of at least one superpower, for questioning Sweden's actual capacity to remain autonomous during a future oil crisis. Clearly, if this is so, it is a conclusion which goes against the very grain of Swedish foreign policy. As the 1978 Parliamentary Defence Committee has phrased it, Swedish policy 'must create neither distrust nor expectations on the part of the superpowers. . . . A relationship of too great a dependency can . . . be utilized to demand political or military favours'.[67] In other words, in so far as the claim is valid that within the IEA framework 'nothing important can happen unless the United States desires it', does it not follow that to the extent that Swedish energy policy during a crisis would be tied to IEA decisions, it is at least to some degree beholden to American energy policy as well—and thus, by extension, also to American foreign policy actions?[68] The mere suspicion that this conclusion may be entertained by either or both superpowers is, I here contend, reason enough for the present Swedish government to give its energy policy renewed and serious consideration.

CONCLUSION

Nils Andrén, the doyen of Swedish security policy analysts, has written the following apropos our subject-matter:

> The military strength of a country may consist of its own resources, of borrowed strength, or of a combination of both. The logic behind not-alignment and neutrality is that Sweden cannot have access to borrowed strength as a peacetime dissuasion. It has to rely on its own efforts.[69]

I have argued that precisely the same applies to non-military dimensions of national security, such as access at an affordable price to commodities—such as oil—which are of vital importance to a nation's welfare and security. Yet on this front, I have also attempted to show, Sweden has on the whole tried to live off 'borrowed strength', with consequences for her neutrality which, if the interpretation given here of its logic is substantially correct, must necessarily affect her ability to exercise the type of influence which has hitherto been the mainstay of this policy. Becoming even partially dependent on the IEA undermines the symmetry of the power relationships with the superpowers which defines the special nature of Swedish neutrality. Avoiding all such impediments to the pursuit of this policy during peacetime thus constitutes a fundamental structural requirement of the logic of precommitment which has constituted a *sine qua non* for the cognitive credibility of this policy since its inception almost four decades ago. While pursuing domestic energy policies which have been unusually responsive to both exogenous contingencies and internal political pressures, Sweden has not, in my view, been equally mindful of this critical requirement.

NOTES

1. Henry R. Nau, 'The Evolution of US Foreign Policy in Energy: From Alliance Politics to Politics-as-Usual', in Robert M. Lawrence and Martin O. Heisler (eds.), *International Energy Policy* (Lexington, 1980), p. 37.

2. See the discussion in Chapter 2, p. 11.
3. See Hanns Maull's discussion of the dimensions of vulnerability in his 'Managing Energy Insecurity', in Joan Pearce (ed.), *The Third Oil Shock* (London, 1983), pp. 90f.
4. Melvin A. Conant, *The Oil Factor in U.S. Foreign Policy* (Lexington, 1982), p. xv.
5. Alvin L. Alm, 'Introduction: Managing Oil Shocks', in Alvin L. Alm and Robert J. Weiner (eds.), *Oil Shock: Policy Response and Implementation* (Cambridge, Mass., 1984), p. 2.
6. Edward L. Morse, 'An Overview: Gains, Costs and Dilemmas', in Pearce, p. 21.
7. Maull, p. 93.
8. Don E. Kash and Robert W. Rycroft, *U.S. Energy Policy: Crisis and Complacency* (Norman, 1984), p. 298.
9. Douglas R. Bohi and William B. Quandt, *Energy Security in the 1980s: Economic and Political Perspectives* (Washington, DC, 1984), p. 34.
10. As noted in a recent report from a discussion at the American Enterprise Institute, the OPEC countries of the Middle East possess 400 billion barrels of oil, while the United States has only 28 billion barrels. Furthermore, while OPEC averages 3,000 barrels per day per well, the United States averages 14 barrels per well per day. See the spring 1987 issue of the *Memorandum* from the American Enterprise Institute for Public Policy Research, p. 12.
11. Joseph S. Nye, 'Energy and Security in the 1980s', *World Politics*, vol. 35, no. 1 (October 1982), p. 133.
12. See above *Memorandum*, p. 12.
13. Nye, p. 133.
14. Morse, p. 24.
15. Ibid. See also pp. 23ff.
16. Ulf Lantzke, 'The Role of International Cooperation', in Alm and Weiner, p. 93.
17. On the nature and consequences of cyclical fluctuations, see Paul Kemezis and Ernest J. Wilson III, *The Decade of Energy Policy: Policy Analysis in Oil-Importing Countries* (New York, 1984), passim; and Morse, pp. 3ff; Maull, pp. 90ff; and Bohi and Quand, pp. 3ff.
18. Conant, p. xvf; see also Morse, pp. 4ff.
19. Kemezis and Wilson, p. viii.
20. Lantzke in Alm and Weiner, p. 79.
21. See, for example, Conant, pp. 75f; Öystein Noreng, 'The International Petroleum Game and Norway's Dilemma', *Cooperation and Conflict*, vol. 17, no. 2 (June 1982), p. 88; William W. Hogan, 'Politics for Oil Importers', in James M. Griffin and David J. Teece (eds.), *OPEC Behaviour and World Oil Prices* (London, 1982), p. 188; and note 52 in Ch. 2.
22. G. John Ikenberry, 'The State and Strategies of International Adjustment', *World Politics*, vol. 39, no. 1 (October 1986), p. 73.
23. Alm, p. 28.
24. Edward N. Krapels, *Oil Crisis Management* (Baltimore, 1980), p. 108.
25. Krapels, p. 108. It should here be added that this is not an aspect which has been wholly ignored in official Swedish discussions of the role of the IEA. See, for example, *Proposition 1983/84:110 Om vissa oljefrågor*, p. 34.
26. See Russell Hardin, *Collective Action* (Baltimore, 1982), pp. 17f.
27. Hardin, p. 20. The classic discussion of the free-rider problem is, of course,

Mancur Olson's *The Logic of Collective Action* (Cambridge, 1965, reprinted with a new appendix 1971). Although Olson has applied this logic specifically to the analysis of unions, professional associations, industry lobbies and farm lobbies, it is equally applicable to all other associations premised on the provision of collective goods. Thus, for example, it has been utilized by Charles K. Kindleberger and Susan Strange in their studies of the free trade system—in their view, a public good—and its built-in tendencies toward free-riding in the form of protective trade policies.

28. Hardin, p. 31.
29. John A.C. Conybeare, 'Public Goods, Prisoners' Dilemmas and the International Political Economy', *International Studies Quarterly*, vol. 28, no. 1 (March 1984), pp. 19ff.
30. Conybeare, p. 20.
31. In this connection see, for example, Glenn H. Snyder and Paul Diesing, *Conflict among Nations: Bargaining, Decision Making, and System Structure in International Crises* (Princeton, 1977), p. 195ff.
32. See Olson, Ch. 6; Hardin, pp. 32ff; and the critique in Brian Barry, *Sociologists, Economists and Democracy* (Chicago, 1978), pp. 37ff.
33. Thus Robert O. Keohane has written that although the IEA 'has relatively strong formal powers, as international organizations go, and has developed elaborate rules, it rarely implements those rules in ways that limit the autonomy either of governments or of those powerful transnational actors, the international oil companies. When the agency is pressed to do so, the exercise of rule-making and compliance tends to become symbolic'. Robert O. Keohane, *After Hegemony: Cooperation and Discord in the World Political Economy* (Princeton, 1984), p. 237.
34. Conybeare, p. 6.
35. Ibid. See also the game matrix in Hardin, p. 26.
36. See Conybeare, p. 20.
37. On co-operative strategies in Prisoners' Dilemmas, see, for example, Robert Axelrod, *The Evolution of Cooperation* (New York, 1984). In so far as they usually build on the efficacy of iterated games, such strategies will tend to be inappropriate in the face of the infrequent and unexpected crises discussed here. For a recent critique of approaches of this kind, see Deborah Welch Larson, 'Game Theory and the Psychology of Reciprocity', paper prepared for the annual meeting of the American Political Science Association, Chicago, 3–6 September 1987.
38. Daniel Badger and Robert Belgrave, *Oil Supply and Price: What Went Right in 1980?* (London, 1982), p. 111.
39. See Keohane, pp. 226 (fn. 5) and 229ff (especially fn. 8); and Badger and Belgrave, pp. 112f.
40. Keohane, pp. 229f.
41. Ibid., p. 230.
42. Ibid., pp. 230f.
43. Alm, p. 26. Although the first claim refers to the United States, the rest of the passage is of general applicability.
44. See David A. Baldwin, *Economic Statecraft* (Princeton, 1985), p. 127.
45. See Krapels, pp. 100ff, and Alm, p. 17.
46. See Gunnar Sjöstedt, *Sverige—litet och svagt?* (Stockholm, 1983), pp. 29ff.
47. Edward L. Morse, 'An Overview: Gains, Costs and Dilemmas', in Joan Pearce

(ed.), *The Third Oil Shock: The Effects of Lower Oil Prices* (London, 1983), p. 24.
48. Morse, p. 24.
49. See *Prop. 1983/84:110*.
50. Wilhelm Agrell, *Sveriges civila säkerhet* (Stockholm, 1984), p. 16. My translation.
51. Rolf H. Lindholm, *Sveriges neutralitet* (Lund, 1987), p. 46.
52. Louis Turner, *UK Interests and the International Energy Agency* (London, 1982), p. 2. My emphases.
53. See ibid., pp. 10f.
54. Ibid., p. 9.
55. *Regeringens proposition 1975:20 Energihushållning m.m.*, p. 22. My translation.
56. The definitive story of these events is Ingemar Dörfer's intriguing narration in *Arms Deal: The Selling of the F-16* (New York, 1983). See also Lindholm, pp. 51ff. It should be added that a reservation to this commitment by the Swedish government was made for the very unlikely event—given the veto powers of the US—that the Security Council of the United Nations would decide to prohibit the sale of war material to these countries.
57. Lindholm, p. 53.
58. Ibid.
59. Nils Andrén, 'Changing Strategic Perspectives in Northern Europe', in Bengt Sundelius (ed.), *Foreign Policies of Northern Europe* (Boulder, 1982), p. 104.
60. Quoted in Nau, p. 37.
61. Melvin A. Conant, *The Oil Factor in US Foreign Policy 1980–1990* (Lexington, 1982), p. 83.
62. Krister Wahlbäck, 'The Nordic Region in Twentieth-Century European Politics', in Sundelius, p. 35.
63. Kemezis and Wilson pp. 26f.
64. Leon N. Lindberg, 'The Energy Crisis: A Political Economy Perspectice', in Leon N. Lindberg (ed.), *The Energy Syndrome: Comparing National Responses to the Energy Crisis* (Lexington, 1977), p. 21.
65. David A. Baldwin, 'Interdependence and Power: A Conceptual Analysis', *International Organization*, vol. 34 (Autumn 1980), p. 489.
66. See, in particular, Wilhelm Agrell's conclusions in *Spelet om neutraliteten* (Stockholm, 1983), pp. 36ff.
67. *SOU 1979:42 Vår säkerhetspolitik*, pp. 107f.
68. This claim is made by Conant, p. 74.
69. Andrén, p. 90.

Bibliography

Agrell, Wilhelm (1983). *Spelet om neutraliteten*. Stockholm.

(1984). *Sveriges civila säkerhet*. Stockholm.

(1985). 'Har förutsättningarna för neutralitetspolitiken förändrats?' in *Tio debattinlägg om svensk säkerhetspolitik*. Stockholm.

(1986). *Bakom ubåtskrisen*. Stockholm.

(1986). 'Behind the Submarine Crisis: Evolution of the Swedish Defence Doctrine and Soviet War Planning', *Cooperation and Conflict*, vol. 21.

Alm, Alvin L. (1984). 'Introduction: Managing Oil Shocks', in Alvin L. Alm and Robert J. Weiner (eds.), *Oil Shock: Policy Response and Implementation*. Cambridge, Mass.

Anderson, Charles W. (1987). 'Political Philosophy, Practical Reasons, and Policy Analysis', in Frank Fischer and John Forester (eds.), *Confronting Values in Policy Analysis: The Politics of Criteria*. Newbury Park.

Andrén, Nils (1967). *Power-Balance and Non-Alignment*. Stockholm.

(1982). 'Changing Strategic Perspectives in Northern Europe', in Bengt Sundelius (ed.), *Foreign Policies in Northern Europe*. Boulder.

Ashley, Richard K. (1984). 'The Poverty of Neorealism', in Robert O. Keohane (ed.), *Neorealism and Its Critics*. New York.

Åström, Sverker (1983). *Sweden's Policy of Neutrality*. Stockholm.

Axelrod, Robert (1984). *The Evolution of Cooperation*. New York.

Badger, Daniel, and Belgrave, Robert (1982). *Oil Supply and Price: What Went Right in 1980?* London.

Baldwin, David A. (1979). 'Power Analysis and World Politics: New Trends versus Old Tendencies', *World Politics*, vol. 31.

(1980). 'Interdependence and Power: A Conceptual Analysis', *International Organization*, vol. 34.

(1985). *Economic Statecraft*. Princeton.

Barry, Brian (1978). *Sociologists, Economists and Democracy*. Chicago.

Beck, Robert J. and Jean, Paul F. (1986). 'American Hegemonic Leadership and the IEA', paper presented to the American Political Science Association convention, Washington, DC, August 1986.

Bohi, Douglas R. and Quandt, William B. (1984). *Energy Security in the 1980s: Economic and Political Perspectives*. Washington, DC.

Brodin, Katarina (1977). *Studiet av utrikespolitiska doktriner*. Stockholm.

Bull-Berg, Hans Jacob (1987). *American International Oil Policy: Causal Factors and Impact*. London.

Buzan, Barry (1985). *People, States and Fear: The National Security Problem in International Relations*. Brighton.

Carlgren, Wilhelm (1973). *Svensk utrikespolitik 1939–1945*. Stockholm.

Carlsnaes, Walter (1981). 'Foreign Policy and the Democratic Process', *Scandinavian Political Studies*, vol. 4.

(1986). *Ideology and Foreign Policy: Problems of Comparative Conceptualization*. Oxford.

Conant, Melvin A. (1982). *The Oil Factor in US Foreign Policy, 1980–1990*. Lexington.

Conybeare, John A.C. (1984). 'Public Goods, Prisoners' Dilemmas and the International Political Economy', *International Studies Quarterly*, vol. 28.

Dahl, Robert (1976). *Modern Political Analysis*. Englewood Cliffs, NJ.

Deese, David A. and Miller, Linda B. (1981). 'Western Europe', in David A. Deese and Joseph S. Nye, Jr (eds.), *Energy and Security*. Cambridge, Mass.

DFE Report No. 13 (1978). *Energy Research and Development in Sweden 1978/81*. Stockholm.

No. 49 (1982). *Energy R & D in Sweden, 1981–1984*. Stockholm.

Ds I 1984:10. *Olja och kol: Förslag till lag om beredskapslagring*. Stockholm.

1984:13. *Energiberedskap för kristid, Appendix 2 (Översyn olja 83)*. Stockholm.

Dörfer, Ingemar (1983). *Arms Deal: The Selling of the F-16*. New York.

Ebinger, Charles K. (1982). *The Critical Link: Energy and National Security in the 1980s*. Cambridge, Mass.

Elster, Jon (1979). *Ulysses and the Sirens: Studies in Rationality and Irrationality*. Cambridge.

(1983). *Explaining Technical Change*. Cambridge.

(1983). *Sour Grapes: Studies in the Subversion of Rationality*. Cambridge.

Gately, Dermot (1982). 'OPEC and the Buying-Power Wedge', in James Plummer (ed.), *Energy Vulnerability*. Cambridge, Mass.

Goldmann, Kjell (1978). *Det internationella systemet*. Stockholm.

Goodman, Gordon T., Kristoferson, Lars A. and Hollander, Jack M. (eds.) (1981). *The European Transition from Oil*. London.

Hardin, Russell (1982). *Collective Action*. Baltimore.

Haskel, Barbara G. (1976). *The Scandinavian Option*. Oslo.

Hogan, William W. (1982). 'Politics for Oil Importers', in James M. Griffin and David J. Teece (eds.), *OPEC Behaviour and World Oil Prices*. London.

Holmberg, Sören and Asp, Kent (1984). *Kampen om kärnkraften*. Stockholm.

Holmström, Barry (1981). 'Realism, moralism och neutralitetspolitik: Östen Undén som anonym ledarskribent', *Statsvetenskaplig tidskrift*, no. 2.

Holsti, K.J. (1986). 'Politics in Command: Foreign Trade as National Security Policy', *International Organization*, vol. 40.

Hughes, Barry B. (1986). 'The First Two Oil Shocks: Policy Response and Effectiveness', *Policy Studies Review*, vol. 5.

Huldt, Bo (1986). 'Svensk nedrustnings- och säkerhetspolitik från tjugotalet till åttiotalet', in Bo Hugemark (ed.), *Neutralitet och försvar: Perspektiv på svensk säkerhetspolitik 1809–1985*. Stockholm.

Höll, O. (ed.) (1983). *Small States in Europe and Dependence*. Vienna.

Ikenberry, G. John (1986). 'The State and Strategies of International Adjustment', *World Politics*, vol. 39.

(1986). 'The Irony of State Strength: Comparative Responses to the Oil Shocks in the 1970s', *International Organization*, vol. 40.

Janos, Andrew C. (1986). *Politics and Paradigms: Changing Theories of Change in Social Science*. Stanford.

Jervas, Gunnar (1986). 'Sweden in a Less Benign Environment', in Bengt Sundelius (ed.), *The Neutral Democracies and the New Cold War*. Boulder.

Johansson, Alf (1984). *Per Albin och kriget*. Stockholm.

Johansson, Alf and Norman, Torbjörn (1986). 'Den svenska neutralitetspolitiken i historiskt perspektiv', in Bo Hugemark (ed.), *Neutralitet och försvar: Perspektiv på svensk säkerhetspolitik 1809–1985*. Stockholm.

Johansson, Thomas B. et al. (1983). 'Sweden Beyond Oil: The Efficient Use of Energy', *Science*, vol. 219.

Kash, Don E. and Rycroft, Robert W. (1984). *US Energy Policy: Crisis and Complacency*. Norman.

Kemezis, Paul, and Wilson Ernest J., III (1984). *The Decade of Energy Policy: Policy Analysis in Oil-Importing Countries*. New York.

Kemp, Geoffrey (1981). 'Military Force and Middle East Oil', in David A. Deese and Joseph S. Nye, Jr, (eds.), *Energy and Security*. Cambridge, Mass.

Keohane, Robert O. (1984). *Beyond Hegemony: Cooperation and Discord in the World Political Economy*. Princeton.

Keohane, Robert O. and Nye, Jr, Joseph S. (1977). *Power and Interdependence*. Boston.

Knorr, Klaus (1977). 'Economic Interdependence and National Security', in Klaus Knorr and Frank M. Tragle (eds.), *Economic Issues and National Security*. Lawrence, Kansas.

Koblik, Steven (1972). *Sweden: The Neutral Victor*. Lund.

Kohl, Wilfred L. (1982). 'Introduction: The Second Oil Crisis and the Western Energy Problem', in Wilfred L. Kohl (ed.), *After the Second Oil Crisis: Energy Policies in Europe, America, and Japan*. Lexington, Mass.

Krapels, Edward N. (1980). *Oil Crisis Management: Strategic Stockpiling for International Security*. Baltimore.

Lantzke, Ulf (1984). 'The Role of International Cooperation', in Alvin L. Alm and Robert J. Weiner (eds.), *Oil Shock: Policy Response and Implementation*. Cambridge, Mass.

Larsson, Sven-Erik (1986). *Regera i koalition*. Stockholm.

Lasswell, Harold, and Kaplan, Abraham (1950). *Power and Society*. New Haven.

Levy, Walter (1978/79). 'Oil Policy and OPEC Development', *Foreign Affairs*, vol. 57.

Lieber, Robert J. (1982). 'Cohesion and Disruption in the Western Alliance', in Daniel Yergin and Martin Hillenbrand, (eds.), *Global Insecurity: A Strategy for Energy and Economic Renewal*. Boston.

(1982). 'Energy and Foreign Policy: Western Responses to the Oil Crises', paper presented to the XII World Congress of the International Political Science Association, Rio de Janeiro, August 1982.

(1983). *The Oil Decade: Conflict and Cooperation in the West*. New York.

Lindberg, Leon N. (1977). 'The Energy Crisis: A Political Economy Perspective', in Leon N. Lindberg (ed.), *The Energy Syndrome*. Lexington.

Lindell, Ulf, and Persson, Stefan (1986). 'The Paradox of Weak State Power: A Research and Literature Overview', *Cooperation and Conflict*, vol. 21.

Lindholm, Rolf H. (1987). *Sveriges neutralitet*. Lund.

Lowi, Theodore S. (1964). 'American Business, Public Policy, Case Studies and Political Theory', *World Politics*, vol. 16.

(1972). 'Four Systems of Policy, Politics, and Choice', *Public Administrative Quarterly*, vol. 32.
Lucas, Nigel (1982). *Western European Energy Policies: A Comparative Study*. Oxford.
Lukes, Steven (1977). *Essays in Social Theory*. London.
Lundestad, Geir (1980). *America, Scandinavia, and the Cold War, 1945–1949*. Oslo.
Lundgren, L. (1978). *Energipolitik i Sverige 1890–1975*. Stockholm.
Lönnroth, M., Johansson, T.B. and Steen, P. (1980). *Solar versus Nuclear: Choosing Energy Futures*. Oxford.
(1980). *Energy in Transition*. Berkeley.
Maoz, Zeev, and Felsenthal, Dan S. (1987). 'Self-Binding Commitments, the Inducement of Trust, Social Choice, and the Theory of International Cooperation', *International Studies Quarterly*, vol. 31.
Maull, Hanns (1981). 'Time Wasted: The Politics of European Energy Transition', in Gordon T. Goodman, Lars A. Kristoferson and Jack M. Hollander (eds.), *The European Transition from Oil*. London.
(1983). 'Managing Energy Insecurity', in Joan Pearce (ed.), *The Third Oil Shock*. London.
Mork, Knut Anton (1982). 'The Economic Cost of Oil Supply Disruptions', in James Plummer (ed.), *Energy Vulnerability*. Cambridge, Mass.
Morse, Edward L. (1983). 'An Overview: Gains, Costs and Dilemmas', in Joan Pearce (ed.), *The Third Oil Shock*. London.
Möller, Yngve (1986). *Östen Undén: En biografi*. Stockholm.
Nau, Henry (1980). 'The Evolution of US Foreign Policy in Energy: From Alliance Politics to Politics-as-Usual', in Robert M. Lawrence and Martin O. Heisler (eds.), *International Energy Policy*. Lexington.
Noreen, Erik (1983). 'The Nordic Balance: A Security Policy Concept in Theory and Practice', *Cooperation and Conflict*, vol. 18.
Noreng, Oystein (1981). 'State Oil Trading and the Perspective of Shortage', in Paul Tempest (ed.), *International Energy Options*. Cambridge, Mass.
(1982). 'The International Petroleum Game and Norway's Dilemma', *Cooperation and Conflict*, vol. 18.
Nye, Joseph S. (1981). 'Energy and Security', in David A. Deese and Joseph S. Nye, Jr (eds.), *Energy and Security*. Cambridge, Mass.
(1982). 'Energy and Security in the 1980s', *World Politics*, vol. 35.
Nye, Joseph S., Deese, David A. and Alm, Alvin L. (1981). 'Conclusion: A US Strategy for Energy Security', in David A. Deese and Joseph S. Nye, Jr (eds.), *Energy and Security*. Cambridge, Mass.
Olson, Mancur (1965). *The Logic of Collective Action*. Cambridge.
Oppenheim, Felix (1981). *Political Concepts: A Reconstruction*. Oxford.
Paris, David C. and Reynolds, James F. (1983). *The Logic of Policy Inquiry*. New York.
Pindyck, Robert S. and Rothenburg, Julio J. (1984). 'Energy Shocks and the Macroeconomy', in Alvin L. Alm and Robert J. Weiner (eds.), *Oil Shock: Policy Response and Implementation*. Cambridge, Mass.
Plummer, James (1982). 'Conclusions: Analytical Results, Directions for Policy, and Important Unanswered Questions', in James Plummer (ed.), *Energy Vulnerability*. Cambridge, Mass.
(1982). 'Energy Vulnerability Policy: Making It Through the Next Few Years

without Enormous Losses', in James Plummer (ed.), *Energy Vulnerability*. Cambridge, Mass.

Plummer, James, Chao, Hung-Po, Gately, D. and Gilbert, Richard (1982). 'Basic Concepts, Assumptions, and Numerical Results', in James Plummer (ed.), *Energy Vulnerability*. Cambridge, Mass.

Plummer, James, and Weyant, John P. (1982). 'International Institutional Approaches to Energy Vulnerability Problems', in James Plummer (ed.), *Energy Vulnerability*. Cambridge, Mass.

Popper, Karl R. (1963). *Conjectures and Refutations*. London.

Ray, G.F. (1982). 'Europe's Farewell to Full Employment', in Daniel Yergin and Martin Hillenbrand (eds.), *Global Insecurity: A Strategy for Energy and Economic Renewal*. Boston.

Regeringens proposition 1975:30. *Energihushållning m.m.* Stockholm.

1976/77:53. *Med förslag till lag om särskilt tillstånd att tillföra kärnreaktor kärnbränsle, m.m.* Stockholm.

1977/78:76. *Med energisparplan för befintlig bebyggelse*. Stockholm.

1978/79:115. *Om riktlinjer för energipolitiken*. Stockholm.

1980/81. *Bilagsdel*. Stockholm.

1983/84:110. *Om vissa oljefrågor*. Stockholm.

1984/85:120. *Riktlinjer för energipolitiken*. Stockholm.

Rowen, Henry S. and Weyant, John P. (1982). 'Trade-Offs Between Military Policies and Vulnerability Policies', in James Plummer (ed.), *Energy Vulnerability*. Cambridge, Mass.

Russell, Bertrand (1975). *Power*. London.

Schelling, Thomas C. (1980). 'The Intimate Struggle for Self-Command', *Public Interest*, vol. 60.

Sjöstedt, Gunnar (1983). *Sverige—litet och svagt?* Stockholm.

Snyder, Glenn H. and Diesing, Paul (1977). *Conflict among Nations: Bargaining, Decision Making, and System Structure in International Crises*. Princeton.

SOU 1951:32. *Bränsle och Kraft*. Stockholm.

1956:11. *Atomenergien*. Stockholm.

1956:46. *Bränsleförsörjning i atomåldern*. Stockholm.

1974:72. *Energiforskning: Program för forskning och utveckling*. Stockholm.

1977:56. *Energi—Program för forskning, utveckling och demonstration* (EFUD). Stockholm.

1979:42. *Vår säkerhetspolitik*. Stockholm.

1980:41. *Olja för kristid*. Stockholm.

1986:42. *Bränsle- och drivmedelsberedskapen under kriser och i krig*. Stockholm.

Stobaugh, Robert, and Yergin, Daniel (1980). 'The End of Easy Oil', in Robert Stobaugh and Daniel Yergin (eds.), *Energy Future*. New York.

Sundelius, Bengt (1984). 'Neutralitetspolitikens obesvarade frågor', *Internationella Studier*, no. 4.

(1986). 'Dilemmas and Security Strategies for the Neutral Democracies', in Bengt Sundelius (ed.), *The Neutral Democracies and the New Cold War*. Boulder.

Svensson, Jan-Erik, Hakkarainen, Elsa, and Eriksson, Anders (1982). *Kunskap om energi: En inventering av svenska energisystemstudier och några hypoteser om forskningsbehov*. Stockholm.

Theutenberg, Bo Johnson (1986). *Folkrätt och säkerhetspolitik*. Stockholm.

Turner, Louis (1982). *UK Interests and the International Energy Agency*. London.
Ullman, Richard H. (1983). 'Redefining Security', *International Security*, Vol. 88.
Vedung, Evert (1979). *Kärnkraften och regeringen Fälldins fall*. Stockholm.
(1980). 'Kärnkraften och partisystemet i Sverige', *Politica*, vol. 12.
(1982). *Energipolitiska utvärderingar 1973–81*. Stockholm.
(1984). 'Main Trends in Swedish Energy Policy in Response to the Oil Crisis', paper presented to a workshop of the European Consortium for Political Research, Salzburg, 13–18 April.
Venturi, Robert (1966). *Complexity and Contradiction in Architecture*. New York.
Wahlbäck, Krister (1982). 'The Nordic Region in Twentieth-Century European Politics', in Bengt Sundelius (ed.), *Foreign Policies in Northern Europe*. Boulder.
(1984). *Den svenska neutralitetens rötter*. Stockholm.
Walker, Stephen G. (1987). 'Role Theory and the Origins of Foreign Policy', in Charles F. Hermann, Charles W. Kegley, Jr, and James N. Rosenau (eds.), *New Directions in Foreign Policy*. Boston.
Wallace, William (1986). 'What Price Independence: Sovereignty and Interdependence in British Politics', *International Affairs*, no. 3.
Waltz, Kenneth (1970). 'The Myth of National Interdependence', in Charles Kindleberger (ed.), *The International Corporation*. Cambridge, Mass.
(1979). *Theory of International Politics*. Reading, Mass.
Wildavsky, Aaron, and Tenenbaum, Ellen (1981). *The Politics of Mistrust*. Beverly Hills.
Wilson, Ernest J., III (1984). 'US International Energy Policy: Competing Explanations of Market-Place Interactions', paper given to a workshop of the European Consortium for Political Research in Salzburg, 13–18 April.
(1987). 'World Politics and International Energy Markets', *International Organization*, vol. 41.
Wittrock, Björn and Lindström, Stefan (1984). *De stora programmens tid—Forskning och energi i svensk politik*. Stockholm.
(1985). 'Beyond Organizational Design: Contextuality and the Political Theory of Public Policy', in Kenneth Hanf and Theo A.J. Toonen (eds.), *Policy Implementation in Federal and Unitary Systems*. Dordrecht.
Wittrock, Björn, Lindström, Stefan, and Zetterberg, Kent (1981). 'Implementation beyond Hierarchy: Swedish Energy Research Policy', Stockholm University (mimeo). Stockholm.
Yergin, Daniel (1982). 'Crisis and Adjustment: An Overview', in Daniel Yergin and Martin Hillenbrand (eds.), *Global Insecurity: A Strategy for Energy and Economic Renewal*. Boston.

Index